Emerging Contaminants in Water: Detection and Treatment (Volume 1)

Emerging Contaminants in Water: Detection and Treatment (Volume 1)

Joseph Welker

New York

Published by Syrawood Publishing House,
750 Third Avenue, 9th Floor,
New York, NY 10017, USA
www.syrawoodpublishinghouse.com

Emerging Contaminants in Water: Detection and Treatment (Volume 1)
Joseph Welker

International Standard Book Number: 978-1-64740-378-2 (Hardback)

Cataloging-in-Publication Data

Emerging contaminants in water : detection and
treatment (Volume 1) / Joseph Welker.
p. cm.
Includes bibliographical references and index.
ISBN 978-1-64740-378-2
1. Water--Pollution. 2. Drinking water--Contamination. 3. Water--Purification.
4. Water quality management. I. Welker, Joseph.
TD420 .E44 2023
628.168--dc23

TABLE OF CONTENTS

PREFACE

Emerging contaminants refer to the naturally occurring or synthetic chemicals or substances that have been identified recently and are toxic to human health and environment. These are known as contaminants of emerging concern (CECs). Some major sources of CECs are pharmaceuticals and personal care products, detergents, steroid hormones, industrial chemicals, and pesticides. The complete removal of the CECs requires the use of advanced chemical, physical and biological treatment techniques. Advanced oxidation processes (AOPs) is an emerging chemical technology used for treating wastewater. Ozone, catalyst and UV irradiation are oxidants that combine in AOPs for delivering more efficient wastewater treatment. The CECs have demonstrated low acute toxicity but studies have shown that they may cause significant reproductive effects at low levels of exposure. This book outlines the latest methods for detecting and treating emerging contaminants in water and wastewater in detail. It will help the new readers in foregrounding their knowledge in this area of study.

This book is a result of research of several months to collate the most relevant data in the field.

When I was approached with the idea of this book and the proposal to edit it, I was overwhelmed. It gave me an opportunity to reach out to all those who share a common interest with me in this field. I had 3 main parameters for editing this text:

1. Accuracy - The data and information provided in this book should be up-to-date and valuable to the readers.

2. Structure - The data must be presented in a structured format for easy understanding and better grasping of the readers.

3. Universal Approach - This book not only targets students but also experts and innovators in the field, thus my aim was to present topics which are of use to all.

Thus, it took me a couple of months to finish the editing of this book.

I would like to make a special mention of my publisher who considered me worthy of this opportunity and also supported me throughout the editing process. I would also like to thank the editing team at the back-end who extended their help whenever required.

Joseph Welker

1

Aquatic Environment: Micropollutants and Contaminants

A. S. Stasinakis and G. Gatidou

1.1 INTRODUCTION

Micropollutants are compounds which are found in the μg L^{-1} or ng L^{-1} concentration range in the aquatic environment and are considered to be potential threats to environmental ecosystems. Different groups of compounds are included in this category such as pesticides, PCBs, PAHs, flame retardants, perfluorinated compounds, pharmaceuticals, surfactants and personal care products. Recent studies have indicated the often detection of these compounds in the aquatic environment (Kolpin *et al.*, 2002; Loos *et al.*, 2009). The way that these compounds enter the environment depends on their uses and the mode of application. The major routes seem to be agricultural and urban runoff, municipal and industrial wastewater discharge, sludge disposal and accidental spills (Ashton *et al.*, 2004; Becker *et al.*, 2008; Mompelat *et al.*, 2009).

Once released into the environment, micropollutants are subjected to different processes such as distribution between different phases, biological and abiotic degradation (Halling-Sørensen *et al.,* 1998; Hebberer, 2002a; Birkett and Lester, 2003; Farre *et al.,* 2008). These processes contribute to their elimination and affect their bioavailability. The role of the aforementioned processes in micropollutants' fate depends on the physico-chemical properties of these compounds (polarity, water solubility, vapor pressure) and the type of the environment (natural or mechanical) where the micropollutants are present (groundwater, surface water, sediment, wastewater treatment systems, drinking-water facilities). As a result, different transformation reactions can take place, producing metabolites that often differ in their environmental behavior and ecotoxicological profile from the parent compounds. So far, several effects of these compounds on aquatic organisms have been reported such as acute and chronic toxicity, endocrine disruption, bioaccumulation and biomagnifications (Oaks *et al.,* 2004; Fent *et al.,* 2006; Darbre and Harvey, 2008).

In the next paragraphs, data for the occurrence, fate and effects of some micropollutants' categories will be given. Pesticides have been selected due to their wide use and their well-defined toxicological effects and environmental fate. Five categories of emerging contaminants will be also presented (pharmaceuticals, steroid hormones, perfluorinated compounds, surfactants and personal care products) due to the great interest that has recently arisen for their occurrence in the environment.

1.2 PESTICIDES

The word "pesticide" precisely is referred to an agent that is used to kill an unwanted organism (Rana, 2006). According to EPA (USEPA) pesticide is called an organic compound (or mixture of compounds) which acts against pests (insect, rodent, fungus, weed etc.) by several ways like prevention, destruction, repulse or mitigation. These biologically active chemicals are often called biocides and include several classes such as herbicides, insecticides and fungicides, depending on the type of the pests that they control.

Before 1940, inorganic compounds and few natural agents originated from plants were used as pesticides (Rana, 2006). The great production of synthetic organic compounds for use in pest control was started with the discovery of DDT's insecticidal activity in 1938 and it was continued during and following the Second World War (Matthews, 2006). During the next decades, an exponential increase in production and use of synthetic pesticides was observed worldwide (Rana, 2006). Today, many different classes of pesticides are used, including among others chlorinated hydrocarbons, organophosphoric compounds, substituted ureas and triazines. Despite the positive effects from the usage of synthetic pesticides on public health and global economy, the extensive use of these compounds resulted in serious environmental contamination problems worldwide and deleterious effects on humans and ecosystems. Risks associated with pesticides usage were firstly mentioned by Rachel Carson in her book *Silent Spring*.

1.2.1 Organochlorine insecticides

Insecticides are one of the most significant types of pesticides due to the fact that they can be applied in a short time before harvesting and after crops collection (Manahan, 2004). Initially, insecticides were divided into two main groups: the organochlorines and organophosphates (Matthews, 2006). Both of these groups affect the nervous system of the organisms by inhibiting the enzyme called acetylcholinesterase (Walker *et al.,* 2006).

Organochlorine insecticides are halogenated solid organic compounds, highly lipophilic, with very low water solubility and high persistence. These properties result in remaining of residues in the environment for long time and further accumulation to several animals (Matthews, 2006; Walker *et al.,* 2006). Among the different classes of organochlorine insecticides, dichlorodiphenylethanes, chlorinated cyclodienes (or chlordanes) and hexachlorocyclohexanes are of great concern due to their potential risk to human health and environmental fate (Qiu *et al.,* 2009). The most known dichlorodiphenylethane pesticide is DDT (Figure 1.1). This compound was mainly used for vector control during the Second World War and thereafter was extensively used in agriculture (Walker *et al.,* 2006).

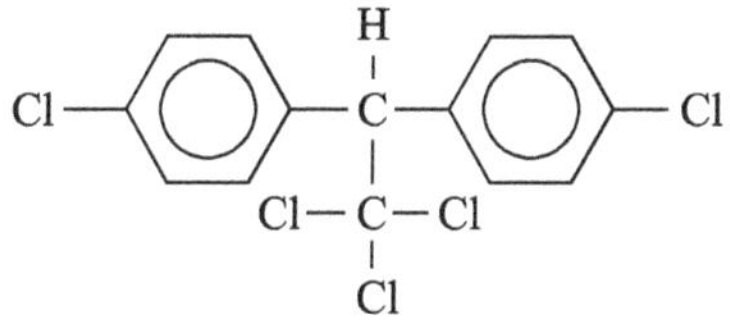

Figure 1.1 The organochlorine insecticide DDT (p′p-dichlorodophenyltrichloroethane)

Chlorinated cyclodienes, such as aldrin, dieldrin or heptachlor (Figure 1.2), were introduced in 1950s and they were used both for crop protection against pests and certain vectors of disease (e.g., tsetse fly) (Walker *et al.,* 2006).

The organochloride insecticides belonging in the group of hexachlorocyclohexanes (HCH) were introduced in the market as a crude mixture of isomers (Walker *et al.,* 2006). Between the five isomers of this mixture, only γ isomer, commonly referred as γ-HCH or lindane (Figure 1.3), found to be effective as insecticide (Manahan, 2004).

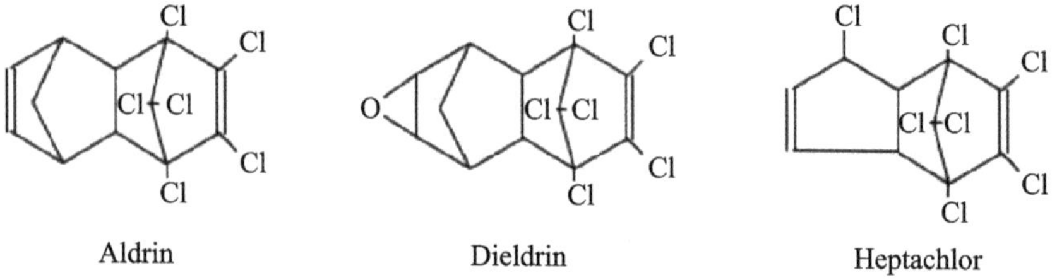

Figure 1.2 The organochloride insecticides aldrin, dieldrin and heptachlor

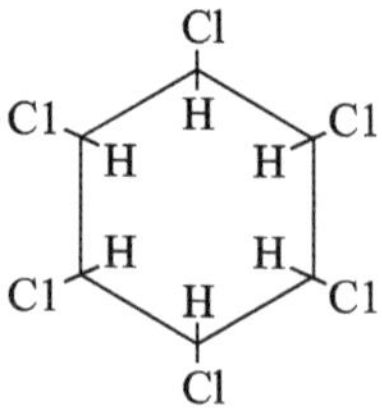

Figure 1.3 The organochloride insecticide lindane (1,2,3,4,5,6-hexachlorocyclo-hexane)

1.2.1.1 Fate

The widespread used organochlorine pesticides present ubiquitous persistence in different environmental media. For instance, a half-life ranging between 3 and 4 years has been estimated for dieldrin in soils, whereas a much higher half-life time (up to 15 years) has been calculated for DDT (UNEP, 2002). These half-lives indicate the the very slow elimination of compounds like DDT by most leaving organisms (IARC/WHO, 1991).

These compounds can enter the aquatic environment in several ways such as run-off from non-point sources or discharge of industrial wastewater. Despite their low water solubility, several organochlorine pesticides are detected worldwide in water column (Table 1.1).

It is entirely known that organochlorine pesticides present high affinity for lipid tissues. As a result, they can be bioaccumulated and biomagnified. According to Zhou *et al.* (2008), log bioconcentration factors (BCFs) of several organochlorine compounds vary from 2.88 to 6.28 for fish, from 3.78 to 6.17 for shrimp and from 3.13 to 5.42 for clams. Through the consumption of aquatic organisms, drinking water and agricultural crops, these compounds reach humans and are excreted to breast milk. Dahmardeh-Behrooz *et al.* (2009) reported mean concentration of total DDTs and HCHs equal to

3563 and 5742 ng g^{-1} (lipid weight), respectively, in human milk in Iran. Besides their banning, these compounds are also detected in human milk in developed countries. Kalantzi *et al.* (2004) reported concentrations as high as 220 and 40 ng g^{-1} (lipid weight) for DDTs and HCHs, resprectively, in human milk from women in England. Moreover, Polder *et al.* (2003) detected even higher concentrations of DDTs (1200 ng g^{-1} lipid weight) and HCHs (320 ng g^{-1} lipid weight) after analyzing human milk in Russia.

Table 1.1 Typical water concentrations of several organochlorine (OC) insecticides

Compound(s)	Concentration (ng L^{-1})	Country	Reference
Total OCs[a]	0.01–9.83	China	Luo *et al.*, 2004
Total OCs	<LOD[b]-112	Greece	Golfinopoulos *et al.*, 2003
Total OCs	0.1–973	China	Zhou *et al.*, 2001
DDT	150–190	India	Shukla *et al.*, 2006
DDT	3.0–33.2	India	Pandit *et al.*, 2002
Lindane	680–1380	India	Shukla *et al.*, 2006
HCH	0.16–15.9	India	Pandit *et al.*, 2002

[a] Referred to the concentration range of compounds: DDTs (DDT and metabolites), aldrin, dieldrin, heptachlor and lindane.
[b] LOD = Limit of Detection.

DDTs or HCH have also been detected in other human matrices such as serum and adipose tissues. Koppen *et al.* (2002) detected concentrations of *p,p*-DDT (2.6 ng g^{-1} fat), *p,p*-DDE (871.3 ng g^{-1} fat) and γ-HCH (5.7 ng g^{-1} fat) in human serum from women in Belgium. Similarly, Botella *et al.* (2004) examined the presence of several organochlorine pesticides in human adipose tissues and blood samples in women from Spain. The detected mean concentrations of total DDTs in the two types of samples were 543.25 ng g^{-1} (human adipose tissues) and 12.10 ng mL^{-1} (blood samples), indicating either a relatively recent exposure or cumulative past exposure.

1.2.1.2 Effects

Toxicity of organochlorine pesticides depends on several parameters including the structure of each compound, the different moieties attached to initial molecule, the nature of substituents (Kaushik and Kaushik, 2007). In many cases, compounds are considered to be moderately toxic to mammals and highly toxic to aquatic organisms. For instance, DDT is considered to be moderately toxic to mammals with LD_{50} values ranging from 113 to 118 mg Kg^{-1} (body weight), while a concentration of 0.6 mg kg^{-1}, reported to cause egg shell thinning for the black duck (UNEP, 2002). Similarly, heptachlor is also considered moderately toxic to mammals but its toxicity to aquatic organisms is high. An LC_{50} value equal to 0.11 μg L^{-1} has been found for pink shrimp (UNEP, 2002). Aldrin and dieldrin are characterized as compounds with high toxicity on aquatic organisms too (Vorkamp *et al.,* 2004). Aldrin's toxicity to aquatic organisms can vary from 1–200 μg L^{-1} for aquatic insects to 2.2–53 μg L^{-1} for fish (96-h LC_{50}). On the contrary, lindane is considered moderately toxic for these organisms. According to UNEP (2002), its estimated LC_{50} values vary from 20 to 90 μg L^{-1} for invertebrates and fish (UNEP, 2002).

Additionally to acute toxicity, organochlorine insecticides are known for their endocrine disrupting effects (Luo *et al.,* 2004). According to Soto *et al.* (1995) and Chen *et al.* (1997) *p,p′*-DDE and *p,p′*-DDT interact with human ERα. Furthermore, *p,p′*-DDE has been reported to act as an antagonist for a human AR (Kelce and Wilson, 1997).

1.2.2 Organophosporous insecticides

Organophosphorous insecticides are synthetic organic compounds which contain phosphorus in their molecule and are organic esters of orthophosphoric acid, phosphonic or phosphorothioic or related acids (Manahan, 2004; Rana, 2006). These compounds were firstly produced for two uses: as insecticides and as chemical warfare gases during Second World War (Walker *et al.,* 2006). Nowadays, most of the organophospates are used as insecticides and their molecules are described with the general formula shown in Figure 1.4.

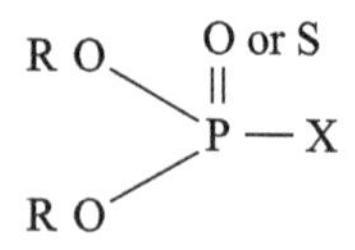

Figure 1.4 The general formula of organophosphorous insecticides. R: alkyl group, X: leaving group

Organophosphates are lipophilic compounds which present higher water solubility and lower stability comparing to organochlorine insecticides. As a result of their lower stability, these pesticides can break down easier by several physicochemical processes leading to shorter remaining times after their release in the environment (Walker *et al.,* 2006). The most commonly used organophosphorus insecticides are phosphorothionates such as methyl parathion and chlorpyrifos (Figure 1.5). These compounds have a sulphur atom (S), instead of oxygen atom (O), connected by double bond with phosphorous atom in their molecule (Manahan, 2004).

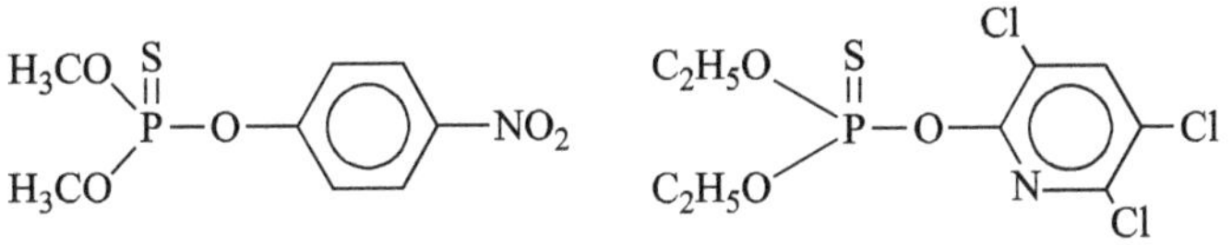

Methyl parathion   Chlorpyrifos

Figure 1.5 The organophosphorous insecticides parathion and chlorpyrifos

1.2.2.1 Fate

Beside the short life of organophosphates in the environment and their low water solubility, these compounds are detected in water column. Some indicative water concentrations of the compounds methyl parathion and chlorpyrifos are presented in Table 1.2.

Table 1.2 Typical water concentrations of the organophosphorous insecticides chlorpyrifos and methyl parathion

Compound	Concentration (ng L^{-1})	Country	Reference
Methyl parathion	<LOD-480	China	Gao *et al.*, 2009
Methyl parathion	<LOD-41	Spain	Claver *et al.*, 2006
Methyl parathion	13–332	Germany	Götz *et al.*, 1998
Methyl parathion	20–270	Spain	Planas *et al.*, 1997
Chlorpyrifos	<LOD-19.41	Italy	Carafa *et al.*, 2007
Chlorpyrifos	<LOD-312	Spain	Claver *et al.*, 2006

LOD: Limit of Detection.

Regarding their fate in the aquatic environment, organophosphorous insecticides can be degraded by oxidation, direct or indirect photodegradation, hydrolysis and adsorption (Pehkonen and Zhang, 2002). All these processes are responsible for the relative short lives of organophosphates in the environment. Araújo *et al.* (2007) investigated the photodegradation of methyl parathion under sunlight and they reported a half life time of about 5 days. Similarly, in another study, Castillo *et al.* (1997) investigated methyl parathion fate in natural waters and they reported half lives of 3 days (groundwater) and 4 days (estuarine and river water). Finally, Wu *et al.* (2006) found that chlorpyrifos is also photodegraded under sunlight and they determined a half life time of about 20 days. Biotic degradation of organophophates can also be occured in the environment. Liu *et al.* (2006) investigated the biodegradation of methyl parathion by two bacteria species *Shewanella* and *Vibrio parahaemolyticus* isolated from river sediments. According to their results, an initial concentration of 50 mg L^{-1} of the compound was almost totally disappeared during one week.

1.2.2.2 Effects

Organophosphates present varying degrees of toxicity in several organisms. For instance, methyl parathion is highly toxic to aquatic organisms and WHO has classified this compound as "extremely hazardous" for the environment, whereas chlorpyrifos is characterized as "moderately hazardous" (WHO, 2004). Furthermore, the transformation of organophosphates may result in the formation of more toxic and persistent metabolites. Dzyadevych *et al.* (2002) reported that methyl paraoxon, a photodegradation product of methyl parathion, found to be at least 10 times more toxic than the parent compound, regarding the inhibition on acetylcholinesterase activity.

Regarding the effects on humans, organophosphates are known for their neurotoxic effects. Additionally, genotoxic effects have been reported. Methyl parathion has been found to produce chromatid exchange in human lymphocytes, while interactions with the double-stranded DNA have also been reported (Rupa *et al.,* 1990; Blasiak *et al.,* 1995). Furthermore, adverse effects of these compounds in reproductive system have been reported in the literature. Salazar-Arredondo *et al.* (2008) investigated human sperm DNA damage of healthy spermatozoa by several organophosphates and their oxon metabolites. According to their results, the tested compounds found to be toxic on sperm DNA, while their metabolites were more toxic then the parent compounds.

1.2.3 Triazine herbicides

The term of herbicide is used to characterize another large group of pesticides which is used against weed control. These compounds act on contact with the plants or are translocated within the plants. According to the time of application, they can be classified to pre- or post-emergence. Furthermore, an herbicide can be a broad spectrum compound or a selective one (Matthews, 2006). Based on their chemical structure, many different groups of herbicides have been reported. Among them, triazines and substituted ureas have significant research interest due to their wide use, persistence and toxic effects.

Triazinic compounds contain three hetorocyclic nitrogen atoms in their molecule (Manahan, 2004). In case that the nitrogen and carbon atoms are interchanged in the ring structure, the herbicide is called symmetric (*s*) triazine. Otherwise, the compound is called non symmetric (*as*) triazine (Figure 1.6). The most widespread and extensively used triazinic herbicide is atrazine. However, there are other members of this group (e.g., simazine) which are also used widely (Strandberg and Scott-Fordsmand, 2002).

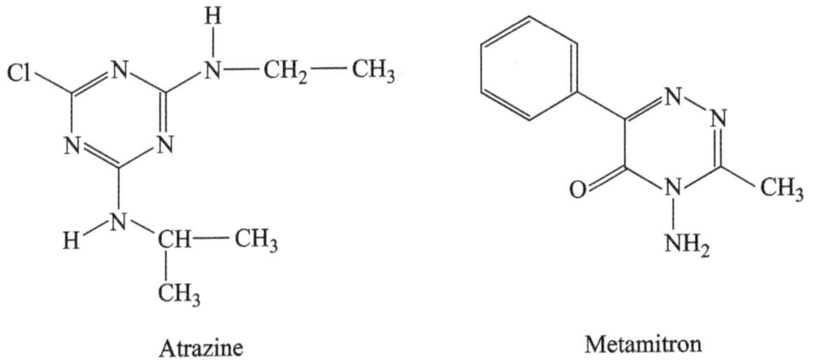

Atrazine

Metamitron

Figure 1.6 Chemical structure of a symmetric (e.g. atrazine) and non symmetric (e.g. metamitron) triazine

Triazinic compounds act as photosystem-II (PSII) inhibitors, affecting photosynthetic electron transport in chloroplasts (Corbet, 1974). Their selectivity is achieved by the inability of target weeds to metabolize and detoxify the herbicidal compound (Manahan, 2004). Triazines are solids, with low vapour pressure at room temperature and varying water solubility, ranging between 5 and 750 mg L^{-1} (Sabik *et al.*, 2000).

1.2.3.1 Fate

Triazines end up to environment via both point (e.g., industrial effluents) and diffuse sources (e.g., agriculture runoff). So far, there are several data regarding their occurrence in the aquatic environment. Some typical water concentrations for two widespread used triazinic herbicides (atrazine and simazine) are presented in Table 1.3.

Triazines are hydrolyzed quickly under acidic or alkaline pH, but at neutral pH are rather stable (Humburg *et al.*, 1989). Photo- and biodegradation of these compounds can also be occurred but s-triazines are found to be more resistant to microbes' attacks. For instance, atrazine is considered as a persistent organic pollutant, with half life ranging between 30 and 100 days (Worthing and Walker, 1987). The aforementioned biotic and abiotic transformations of triazines lead to the formation of metabolites by several mechanisms such as dehalogenation, dezalkylation and deamination (Peñuela and Barceló, 1998).

Table 1.3 Typical water concentrations of atrazine and simazine

Compound	Concentration ($ng\ L^{-1}$)	Country	Reference
Atrazine	1.27–8.18	Italy	Carafa *et al.*, 2007
Atrazine	52–451	Spain	Claver *et al.*, 2006
Atrazine	<LOD-110	Australia	McMahon *et al.*, 2005
Atrazine	<LOD-3870	Greece	Albanis *et al.*, 2004
Atrazine	20–230	Greece	Lambropoulou *et al.*, 2002
Simazine	1.45–25.96	Italy	Carafa *et al.*, 2007
Simazine	49–183	Spain	Claver *et al.*, 2006
Simazine	<LOD-50	Australia	McMahon *et al.*, 2005
Simazine	<LOD-490	Greece	Albanis *et al.*, 2004

LOD: Limit of Detection.

1.2.3.2 Effects

Since triazines have been chemically designed to inhibit photosynthesis and animals lack a photosynthetic mechanism, these compounds are more toxic to plants. Acute toxicity to mammals and birds is low. For instance, atrazine has been classified by WHO (2004) as a compound unlikely to present acute hazard in normal use. Its oral LD_{50} to rats is 3090 mg Kg^{-1} body weight. The compound is slightly toxic to fish and other aquatic organisms and practically nontoxic to birds (UNEP, 2002).

Despite the expected low toxicity to mammals, some triazines have been characterized as potential endocrine disruptors. Atrazine inhibits androgen-mediated development and produces estrogen-like effects in exposed organisms. Furthermore, the occurrence of atrazine in water has been considered responsible for affection of semen quality and fertility in men farmers, as well as increase of breast cancer in women (Fan *et al.*, 2007).

1.2.4 Substituted ureas

The herbicides of this group (e.g., diuron, isoproturon) are derived when hydrogen atoms in the urea molecule are substituted by several chemical groups (Figure 1.7). They have the same biochemical mode of action with triazines, inhibiting photosynthesis (Corbet, 1974).

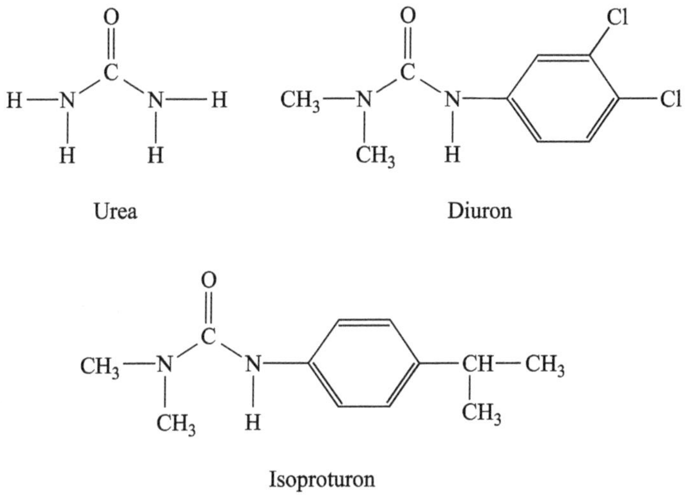

Urea

Diuron

Isoproturon

Figure 1.7 The substituted urea herbicides, diuron and isoproturon

1.2.4.1 Fate

Substituted ureas are transferred to the aquatic environment after their application in crops, via run-off. As a result, they are detected worldwide at concentrations up to few μg L^{-1} (Table 1.4).

Substituted ureas are transformed by both abiotic and biotic processes. Isoproturon, which is a hydrophophic compound, is hydrolysed both in low and high pH values. (Gangwar and Rafiquee, 2007). Salvestrini *et al.* (2002) reported that despite the slow hydrolysis rate of diuron in natural solutions, when this abiotic process takes place is irreversible and the only metabolite is 3,4-dichloroaniline (DCA). Phototransformation of urea herbicides can also be occurred (Shankar *et al.,* 2008). Furthermore, substituted ureas can be subjected to biodegradation and give metabolites which may be more toxic than the parent compound. Goody *et al.* (2002) reported that diuron degradation leads to the formation of the toxic metabolite DCA. In a recent study, Stasinakis *et al.* (2009a) reported that under aerobic and anoxic conditions diuron can be biotransformed to DCA, DCPMU (1-(3,4-dichlorophenyl)-3-methylurea) and DCPU (1-3,4-dichlorophenylurea). Except of these metabolites, a significant part of Diuron seems to be mineralized or/and biotransformed to other unknown compounds.

1.2.4.2 Effects

Similarly to triazines, substituted ureas are expected to be highly toxic in photosynthetic organisms due to their biochemical mode of action. Gatidou and Thomaidis (2007) investigated the toxic effects of diuron on the photosynthetic microorganism *Dunaniella teriolecta* and they estimated an EC_{50} value (after 96h of exposure) of about 6 μg L^{-1}. Fernandez-Alba *et al.* (2002) estimated an EC_{50} value equal to 3.2 μg L^{-1} for seagrass. On the other hand, significant lower toxicity has been reported for crustaceans (8.6 mg L^{-1}, 48 h EC_{50}) and fish (74 mg L^{-1}, 7 day LC_{50}) (Fernandez-Alba *et al.*, 2002).

Table 1.4 Typical water concentrations of the substituted ureas herbicides diuron and isoproturon

Compound	Concentration (ng L^{-1})	Country	Reference
Diuron	<LOD-366	UK	Gatidou *et al.*, 2007
Diuron	7.64–40.78	Italy	Carafa *et al.*, 2007
Diuron	<LOD-105	Spain	Claver *et al.*, 2006
Diuron	30–560	Greece	Gatidou *et al.*, 2005
Diuron	<LOD-80	Australia	McMahon *et al.*, 2005
Diuron	<LOD–3054	Japan	Okamura *et al.*, 2003
Isoproturon	<LOD-92	China	Müller *et al.*, 2008
Isoproturon	0.22–32.08	Italy	Carafa *et al.*, 2007
Isoproturon	LOD<30	Spain	Claver *et al.*, 2006

LOD: Limit of Detection.

In mammals, substituted ureas pose slight toxicity. An oral LD_{50} of diuron in rats equal to 3.4 g Kg^{-1} or a dermal LD_{50} greater than 2 g Kg^{-1} are indicative of its low toxicity in mammals (Giacomazzi and Cochet, 2004). According to WHO (2004), some compounds like isoproturon are classified as slightly hazardous and others such as linuron or diuron as compounds unlikely to present acute hazard in normal use. Regarding ureas' metabolites, DCA has been found to be highly toxic and has been classified as a secondary poisonous substance (Giacomazzi and Cochet, 2004).

1.2.5 Legislation

The excessive usage of synthetic pesticides resulting in contamination problems and harmful effects on humans and ecosystems led several countries to take action concerning their presence in the aquatic environment. European Union with Directive 98/83/EC (EU, 1998) set maximum allowable concentrations in drinking water for individual pesticides (0.1 μg L^{-1}) and for total pesticides (0.5 μg L^{-1}). Furthermore, with the Decision No 2455/2001/EC European

Community established the list of priority substances in the field of water policy, amending Directive 2000/60/EC. Pesticides such as atrazine, simazine, diuron, isoproturon, heptachlor, aldrin, dieldrin, lindane, chlorpyrifos are considered as priority compounds (EU, 2001). Additionally, some European countries have set environmental quality standards (EQSs) for priority compounds. Italy established EQSs for some pesticides in water ranging from 1 ng L^{-1} (chlorpyrifos) to 50 ng L^{-1} (atrazine) (Carafa *et al.*, 2007).

UNEP Governing Council decided in 1997 the immediate international action for the reduction and/or elimination of the emissions and discharges of 12 persistent organic pollutants (POPs) due to their persistent, toxicity and bioaccumulation. This decicion led to the adoption of Stockholm Convention in 2001. Among the different compounds which compose the also known "dirty dozen" are the organochlorine pesticides: aldrin, DDT, dieldrin, heptachlor (UNEP, 2003). According to Master List of Actions Report of UNEP (UNEP, 2003), the above organochlorine pesticides have been banned in most countries worldwide. Furthermore, an International Code of Conduct has been promoting since 1985 by Food and Agriculture Organization of the United Nations. This Code sets standards for governments, pesticide industry and pesticide users (FAO, 2003).

1.3 PHARMACEUTICALS

Pharmaceutically active compounds (pharmaceuticals) are complex molecules with molecular weights ranging from 200 to 500/1000 Da, which are developed and used due to their specific biological activity (Kummerer, 2009). A great number of pharmaceutical compounds (more than 4000 compounds in Europe) are discharged to the environment after human and veterinary usage (Mompelat *et al.*, 2009). In contrast to other micropollutants, that their concentrations will be decreased in the future due to the existed laws and regulations, the use of pharmaceuticals is expected to be increased due to their beneficial health effects.

The first study on human drugs' occurrence in environmental samples appeared in the late 1970s (Hignite and Azarnoff, 1977). The research regarding the effects of these compounds in the environment started in 1990s, when it was discovered that some of these compounds interfere with ecosystems at concentration levels of a few micrograms per liter (Halling-Sørensen *et al.*, 1998). In parallel, during that decade the first optimized analytical methods were developed for the determination of low concentrations of pharmaceuticals in environmental samples (Hirsch *et al.*, 1996; Ternes *et al.*, 1998).

Among pharmaceuticals, non-steroidal anti-inflammatory drugs (NSAIDs), anticonvulsants, lipid regulators and antibiotics are often detected into the aquatic environment and they are considered as a potential group of environmental contaminants. NSAIDs are drugs with analgesic, antipyretic and anti-inflammatory effects. Typical representatives of this category are ibuprofen (IBF, $C_{13}H_{18}O_2$) and diclofenac (DCF, $C_{14}H_{11}C_{l2}NO_2$) (Figure 1.8). Anticonvulsants are used in the treatment of epileptic seizures, with carbamazepine (CBZ, $C_{15}H_{12}N_2O$) being the compound which is often reported in relevant papers. Almost 1000 tons CBZ are estimated to be consumed worldwide (Zhang *et al.*, 2008). Lipid regulators such as gemfibrozil (GEM, $C_{15}H_{22}O_3$) are used to lower lipid levels. Antibiotics are characterized by the great variety of substances such as penicillins, tetracyclines, sulfonamides and fluoroquinolones. In the literature there are available data for all these categories. In this text, data will be given for erythromycin and trimethoprim. Erythromycin ($C_{37}H_{67}NO_{13}$) is a macrolide antibiotic used as human and veterinary medicine, as well as in aquacultures. Trimethoprim (TMP, $C_{14}H_{18}N_4O_3$) is mainly used for treatment of urinary tract infections (Figure 1.8).

Pharmaceuticals are not completely metabolized by the human/animal body. As a result, they are excreted via urine and faeces as unchanged parent compound and as metabolites or conjugates (Heberer, 2002a). Excretion rates are significantly depending on the compound and the mode of application (oral, dermal). Regarding CBZ, after oral administration, almost 28% of the parent compound is discharged through the faeces to the environment, while the rest is absorbed and metabolized by the liver (Zhang *et al.*, 2008). The metabolites of CBZ are excreted with urine. Among them, the most important seem to be 10,11-dihydro-10,11-expoxycarbamazepine (CBZ-epoxide) and trans-10,11-dihydro-10,11-dihydroxycarbamazepine (CBZ-diol) (Reith *et al.*, 2000). Regarding DCF, almost 65% of its oral dosage is excreted through urine (Zhang *et al.*, 2008). The main DCF metabolites detected in urine are $4'$-hydroxy-diclofenac ($4'$-OH-DFC) and $4'$,5-dihydroxy-diclofenac ($4'$-5-diOH-DFC) (Schneider and Degan, 1981). IBF is extensively metabolized in the liver to 2-[4-(2-hydroxy-2-methylpropyl)phenyl]-propionic acid (hydroxyl-IBF) and 2-[4-(carboxypropyl)phenyl]-propionic acid (carboxy-IBF) (Winker *et al.*, 2008). Regarding TMP, almost 80% of the parent compound is excreted, while its main metabolites are 1,3-oxides and $3'$,4-hydroxy derivatives (Kasprzyk-Hordern *et al.*, 2007). Only 5% of erythromycin is excreted unchanged, while its major metabolite is Erythromycin-H_2O (Kasprzyk-Hordern *et al.*, 2007). GEM is metabolized by the liver to four main metabolites and almost 70% of the initial compound is excreted as the glucuronide conjugate in the urine (Zimetbaun *et al.*, 1991).

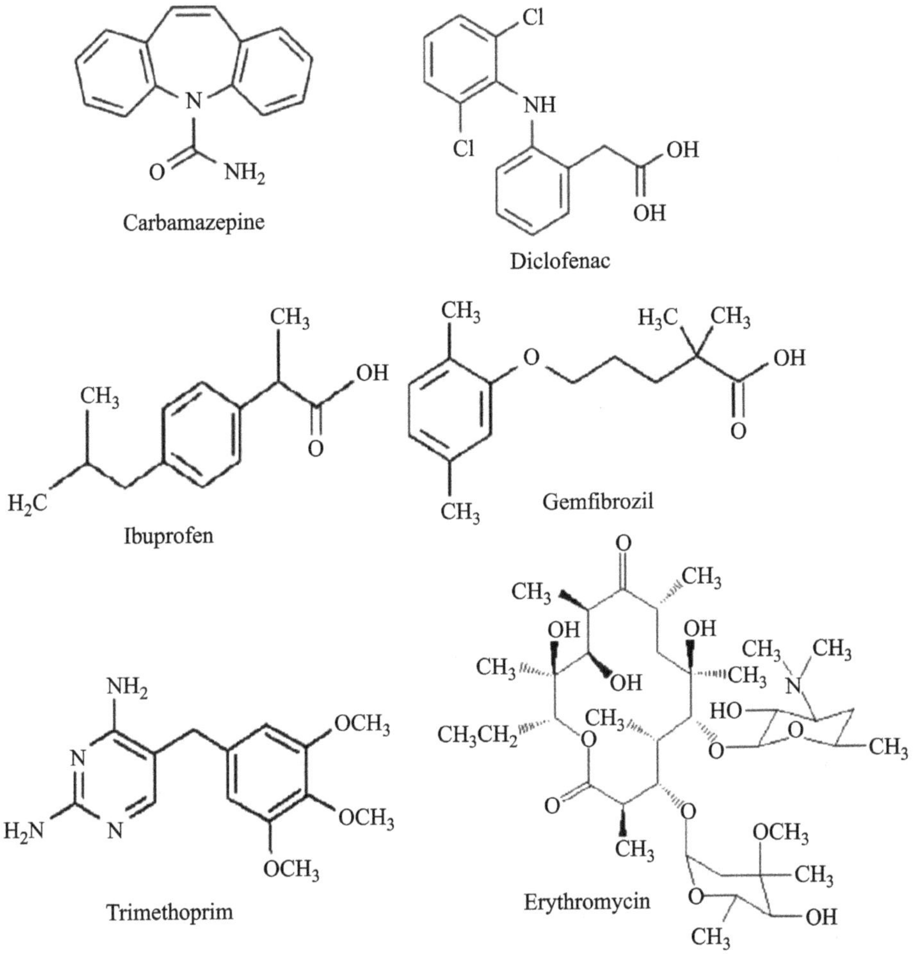

Carbamazepine

Diclofenac

Ibuprofen

Gemfibrozil

Trimethoprim

Erythromycin

Figure 1.8 Chemical structures of selected pharmaceuticals

1.3.1 Fate

Municipal wastewater is the main way for the introduction of human pharmaceuticals and their metabolites in the environment. Moreover, hospital wastewater, wastewater from production industries and landfill leachates may contain significant concentrations of these compounds (Bound *et al.,* 2006; Gomez *et al.,* 2007). Regarding veterinary drugs, they can directly (e.g., use in aquacultures) or indirectly (e.g., manure application and runoff) be released to the environment (Sarmah *et al.,* 2006).

In the literature there are several data regarding removal efficiency of pharmaceuticals in WWTSs. Removal rates are variable between different WWTSs and different compounds (Fent *et al.,* 2006), indicating that they depend on the chemical properties of the compound as well as the treatment process applied. The main mechanisms affecting pharmaceuticals' removal in WWTSs are sorption and biodegradation. NSAIDs and GEM occur as ions at neutral pH and they have little tendency of adsorption to the suspended solids in WWTSs. As a result, they remain in the dissolved phase and they disposed to the environment via the treated wastewater (Fent *et al.,* 2006). On the other hand, basic pharmaceuticals (e.g., fluoroquinolone antibiotics) can be adsorbed to suspended solids and accumulated to the sludge. Regarding the role of biodegradation in WWTSs, this seems to be significant for some compounds (e.g., DCF), while it is of minor importance for others (e.g., CBZ) (Metcalfe *et al.,* 2003; Kreuzinger *et al.,* 2004).

Due to the partial removal of pharmaceuticals during WWTSs, significant concentrations of these compounds are often detected in effluent wastewater, while lower concentrations are detected in surface water and groundwater (Segura *et al.,* 2009; Table 1.5). A recent survey in European river waters revealed that CBZ, DCF, IBF and GEM were detected in 95%, 83%, 62% and 35% of collected samples respectively (Loos *et al.,* 2009). Beside the fact that a significant part of pharmaceuticals are excreted from human/animal bodies as metabolites, so far, in most research papers concentrations of the parent compounds are reported, while there are limited data for the concentrations of their metabolites in the aquatic environment. In a previous study, IBF and its major metabolites hydroxyl- and carboxy-IBF were determined in wastewater and seawater (Weiger *et al.,* 2004). According to the results, hydroxyl-IBF was the major component in treated wastewater (concentrations ranging between 210 to 1130 ng L^{-1}), whereas carboxy-IBF was dominant in seawater samples (concentrations up to 7 ng L^{-1}). The elevated concentrations of hydroxyl-IBF in treated wastewater seem to be due to its excretion from human body, as well as to its formation in activated sludge process (Zwiener *et al.,* 2002). In another study, CBZ and five metabolites were detected in wastewater samples (Miao and Metcalfe, 2003). Among them, 10,11-dihydro-10,11-dihydroxycarbamazepine was detected at much higher concentrations than the parent compound. Moreover, in a recent study, Leclercq *et al.* (2009) reported the presence of six metabolites of CBZ in wastewater samples. Among them, 10,11-dihydro-10,11-trans-dihydroxycarbamazepine was detected at a higher concentration than the parent compound.

Table 1.5 Concentrations of pharmaceuticals in water samples

Substance	Concentration (ng L^{-1})	Country	Reference
Surface Water			
Trimethoprim	<LOD-183	UK	Kasprzyk-Hordern *et al.*, 2008
Erythromycin-H_2O	<LOD-351	UK	Kasprzyk-Hordern *et al.*, 2008
IBF	<LOD-100	UK	Kasprzyk-Hordern *et al.*, 2008
DCF	<LOD-261	UK	Kasprzyk-Hordern *et al.*, 2008
CBZ	<LOD-684	UK	Kasprzyk-Hordern *et al.*, 2008
IBF	<LOD-5044	UK	Ashton *et al.*, 2004
DCF	<LOD-568	UK	Ashton *et al.*, 2004
Erythromycin	<LOD-1022	UK	Ashton *et al.*, 2004
Trimethoprim	<LOD-42	UK	Ashton *et al.*, 2004
Groundwater			
DCF	<LOD-380	Germany	Heberer, 2002b
IBF	<LOD-200	Germany	Heberer, 2002b
GEM	<LOD-340	Germany	Heberer, 2002b
CBZ	<LOD-2.4	USA	Standley *et al.*, 2008
IBF	<LOD-19	USA	Standley *et al.*, 2008
Trimethoprim	1.4–11	USA	Standley *et al.*, 2008
Treated Wastewater			
IBF	20–1820	Europe	Andreozzi *et al.*, 2003
DCF	<LOD-5450	Europe	Andreozzi *et al.*, 2003
CBZ	300–1200	Europe	Andreozzi *et al.*, 2003
Trimethoprim	20–130	Europe	Andreozzi *et al.*, 2003
IBF	780–48240	Spain	Santos *et al.*, 2007
CBZ	<LOD-1290	Spain	Santos *et al.*, 2007
DCF	<LOD	Spain	Santos *et al.*, 2007

(*continued*)

Table 1.5 (*continued*)

Substance	Concentration (ng L^{-1})	Country	Reference
Erythromycin	<LOD-1842	UK	Ashton *et al.*, 2004
Trimethophim	<LOD-1288	UK	Ashton *et al.*, 2004
IBF	240–28000	Spain	Gomez *et al.*, 2007
DCF	140–2200	Spain	Gomez *et al.*, 2007
CBZ	110–230	Spain	Gomez *et al.*, 2007

LOD: Limit of Detection.

Regarding the fate of these compounds to the aquatic environment, they can be adsorbed on suspended solids, colloids and dissolved organic matter or/and undergo biotic, chemical and physico-chemical transformations (Yamamoto *et al.*, 2009). Data on the sorption of pharmaceuticals in sediment and soil have been reported in several studies in the literature (Tolls, 2001; Figueroa *et al.*, 2004; Drillia *et al.*, 2005; Kim and Carlson, 2007). In most of those studies, higher sorption coefficients of pharmaceuticals than those predicted from octanol–water partitioning coefficients (log K_{ow}) were found, suggesting that mechanisms other than hydrophobic partitioning play a significant role in sorption of these compounds (Tolls, 2001). In cases that treated wastewater or sludge are reused for agricultural purposes, highly mobile pharmaceuticals can contaminate groundwater, whereas strongly sorbing compounds can accumulate in the top soil layer (Thiele-Bruhn, 2003). Sorption of pharmaceuticals to soils is affected by the solution chemistry, the type of mineral and organic sorbents and the concentration of dissolved organic matter (DOM) in reused wastewater (Nelson *et al.*, 2007; Blackwell *et al.*, 2007). Experiments with NSAIDs showed that CBZ and DCF can be classified as slow-mobile compounds in soil layers which are rich in organic matter, while their mobility increases significantly in soils which are poor in organic matter (Chefetz *et al.*, 2007).

Pharmaceuticals photodegradation depends on several factors such as the intensity of solar irradiation, the concentration of nitrates, DOM and bicarbonates (Lam and Mabury, 2004). The role of process in pharmaceuticals' fate varies significantly between different compounds of this category (Lam *et al.*, 2004; Benotti and Brownawell, 2009). For instance, phototransformation seems to be the major mechanism of DCF removal in surface water (Buser *et al.*, 1998; Andreozzi *et al.*, 2003). A half-life lower than 1 h has been reported under natural sunlight, while the initial product of DCF photodegradation was 8-chlorocarbazole-1-acetic acid, which is photodegraded even faster than the parent compound (Poiser *et al.*, 2001). In other experiments, Lin and Reinhard (2005) calculated a half-life of GEM equal to 15 hours for river water. On the other hand, CBZ and IBF are photodegraded with much slower rate under sunlight irradiation (Yamamoto *et al.*, 2009). Regarding CBZ, a half-life of 115 hours has been calculated, while 10,11-epoxycarbamazepine was its major phototransformation product (Lam and Mabury, 2005).

So far, most biodegradation studies with pharmaceuticals have focused on their removal during wastewater treatment (Joss *et al.*, 2005; Radjenovic *et al.*, 2009). On the other hand, there are limited data for their biodegradation in the aquatic environment. Lam *et al.* (2004) conducted experiments in a microcosm with several pharmaceuticals and found that photolysis was more important mechanism than biodegradation for CBZ and trimethoprim. Biodegradation experiments with river water showed that IBF and CBZ were relatively stable against microbes (Yamamoto *et al.*, 2009). Half lives of 450–480 h^{-1} and 3000–5600 h^{-1} were calculated for IBF and CBZ, respectively (Yamamoto *et al.*, 2009). In another study, IBF was biodegraded in a river biofilm reactor and its main metabolites were hydroxyl–IBF and carboxy–IBF (Winkler *et al.*, 2001). Experiments with river sediment showed that under aerobic conditions DCF can be biodegraded and its major metabolite was p-benzoquinone imine of 5-hydroxydiclofenac (Groning *et al.*, 2007). Experiments with CBZ and trimethoprim showed that half-lives higher than 40 days were calculated for the biodegradation of these compounds in seawater (Benotti and Brownawell, 2009).

1.3.2 Effects

Experiments with single compounds have shown that acute toxicity of most pharmaceuticals on aquatic organisms seems unlikely for environmental relevant concentrations (Choi *et al.*, 2008; Zhang *et al.*, 2008). Acute effects have been observed at much higher concentrations (100–1000 times) than those usually determined in the aquatic environment (Farre *et al.*, 2008). However, it should be mentioned that pharmaceuticals are usually occurred in the environment as mixtures. Based on the above, several studies have shown that their toxicity to non-target organisms may be occurring at environmentally relevant concentrations due to combined and synergistic effects (Pomati *et al.*, 2008; Quinn *et al.*, 2009). Specifically, toxicity experiments with a mixture of NSAIDs showed that mixture toxicity was found at concentrations at which the single compound showed no or only little effects (Cleuvers, 2004). Moreover, ecotoxicity tests with antibiotics showed that combined toxicity of two antibiotics can lead to either synergistic, antagonistic, or additive effects (Christensen *et al.,* 2006).

On the other hand, chronic toxicity of pharmaceuticals is a matter of great interest, as several aquatic species are exposed to these compounds for their entire life cycle. Beside the above, so far, fewer data are available regarding the long-term effects of pharmaceuticals to aquatic organisms. Schwaiger *et al.* (2004) studied DCF possible effects in rainbow trout after prolonged exposure and they reported histopathological changes of kidney and liver when fish was exposed to 5 $\mu g\ L^{-1}$ DCF for a period of 28 days. In another study, Triebskorn *et al.* (2004) reported that the lowest observed effect concentration (LOEC) for cytological alterations in liver, kidney and gills of rainbow trout was 1 $\mu g\ L^{-1}$ DCF.

Some of the pharmaceuticals seem to bioconcentrate and transport through food chain in other species. Mimeault *et al.* (2005) investigated the uptake of GEM in goldfish and reported that exposure to environmental levels of GEM results to bioconcentration of this compound in plasma. Schwaiger *et al.* (2004) reported that DCF is bioconcentrated mainly in liver and kidney of rainbow trout. Brown *et al.* (2007) reported the bioaccumulation of DCF, IBF and GEM in fish blood of rainbow trout. Several studies have related the presence of DCF residues with decline of vultures' population in India and Pakistan (Oaks *et al.,* 2004; Schultz *et al.,* 2004). Other toxicity effects which have been reported in the literature, include estrogenic activity (Isidori *et al.,* 2009), as well as mutagenic and genotoxic potential of GEM (Isidori *et al.,* 2007). Finally, the release of antibiotics as well as their metabolites into the environment increases the risk of developing bacterial resistance to antibiotics in aquatic ecosystems (Costanzo *et al.,* 2005; Thomas *et al.,* 2005).

1.3.3 Legislation

Despite the great amounts of pharmaceuticals released to the environment, regulations for ecological risk assessment are largely missing. In USA, environmental assessments of veterinary pharmaceuticals are required by the U.S. Food and Drug Administration (FDA) since 1980 (Boxall *et al.,* 2003). Regarding human pharmaceuticals, an environmental assessment report should be provided in cases that the expected concentration of the active ingredient of the pharmaceutical in the aquatic environment is expected to be equal to or higher than 1 $\mu g\ L^{-1}$ (FDA-CDER, 1998). In European Union, the first requirement for ecotoxicity testing was established in 1995 for veterinary pharmaceuticals, according to the European Union Directive 92/18/EEC and the corresponding "Note for Guidance" (EMEA, 1998). During the last decade,

European Commission published Directive 2001/83/EC amended by Directive 2004/27/EC (for human pharmaceuticals) and Directive 2001/82/EC amended by 2004/28/EC (for veterinary pharmaceuticals), indicating that authorization for pharmaceuticals must be accompanied by environmental risk assessment.

There are rare cases where limit values have been set for the presence of pharmaceuticals in the aquatic environment. In such a case, California set water quality standard (19 ng L^{-1}) for lindane (a compound used as pharmaceutical in treatment of head lice) in drinking water sources.

1.4 STEROID HORMONES

Steroid hormones are a group of compounds controlling endocrine and immune system. The major classes of natural hormones are estrogens (e.g., estradiol, estrone, estriol), androgens (e.g., progesterone, androstenedione), progestagents (e.g., progesterone) and corticoids (e.g., cortisol). Several synthetic hormones such as ethinylestradiol, mestranol, dexamethanose have also been produced apart from the aforementioned endogenous hormones.

Among these compounds, estrone (E1), 17β-estradiol (E2), estriol (E3) and ethinylestradiol (EE2) (Figure 1.9) have received more scientific attention since they consider to be the most important contributors to estrogenicity of treated wastewaters and surface waters (Rodgers-Grey *et al.,* 2000). These compounds end up in the environment through wastewater effluents, untreated discharges, runoff of manure and sewage sludge reuse. Aquaculture is another important source of estrogens in the environment (Fent *et al.,* 2006). Fish food additives containing hormones are directly added into the water. Therefore, these compounds can end up in the aquatic environment due to overfeeding or loss of appetite of fish (a phenomenon normally observed in sick organisms).

Steroid hormones are excreted by humans (Länge *et al.,* 2002). Several studies have shown that the gender, the pregnancy or menopause can differentiate the excretion rates of these compounds. For instance, E1 found to have an excretion rate of about 11, 5 and 1194 μg d^{-1} for premenopausal, postmenopausal and pregnant women, respectively, indicating that pregnant women may contribute in a large extent to the total amount of natural estrogens excreted by humans. For men, the excretion rate of E1 estimated to be 3.9 μg d^{-1} (Liu *et al.,* 2009). Natural estrogens in urine are mainly excreted in sulfate or glucuronide conjugates. However, free estrogens have also been detected in feces. Glucuronides were reported to easily change to their free estrogens, whereas sulfates were more resistant to biotransformation (D'Ascenzo *et al.,* 2003).

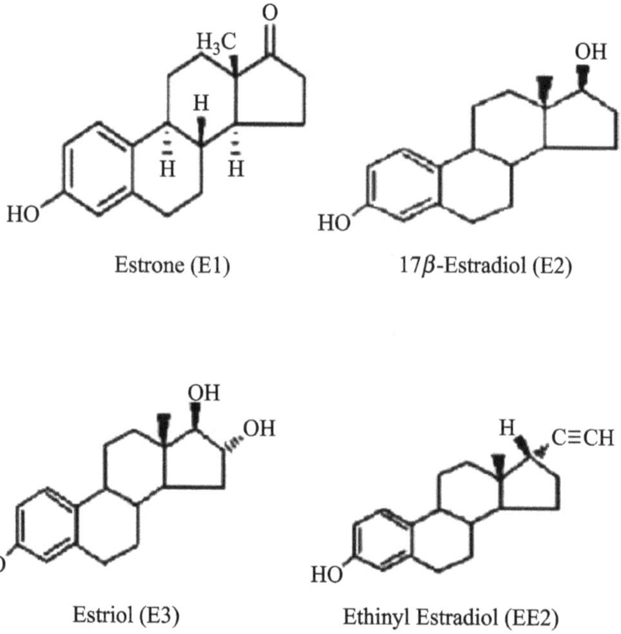

Estrone (E1) 17β-Estradiol (E2)

Estriol (E3) Ethinyl Estradiol (EE2)

Figure 1.9 Molecular formula of estrogens: estrone (E1), 17β-estradiol (E2), estriol (E3) and ethinyl estradiol (EE2)

1.4.1 Fate

Estrogenic compounds have been detected in WWTSs and surface waters. A survey of the US Geological Service indicated that these compounds are often detected in water bodies. Specifically, they were detected at percentages varied from 6% to 21% of the total number of analyzed samples. The median concentrations found to be between 0.03 and 0.16 $\mu g\ L^{-1}$ (Kolpin *et al.*, 2002). Other authors have also reported the presence of steroids in surface and drinking water (Table 1.6). Detection of steroid hormones in drinking water at concentration levels similar to those found in surface waters indicate that these compounds do not totally being removed during water treatment (Ning *et al.*, 2007).

Wastewater seems to be the major transport route of these compounds in the environment. Servos *et al.* (2005) detected considerable effluent estrogenicity, possibly due to hormonally active intermediates of estrogens formed either due to degradation during wastewater treatment or by cleavage of estrogen conjugates (Ning *et al.*, 2007). Some indicative concentrations of E1, E2 and EE2 compounds in treated wastewater are given in Table 1.7.

Table 1.6 Occurrence of steroid hormones in water column

Compound	Concentration (ng L^{-1})	Country	Reference
Estrone (E1)	DW: 0.70	Germany	Kuch and Ballschmiter, 2001
	SW: 1.5–12	Italy	Lagana *et al.*, 2004
	SW: 1.4–1.8	France	Cargouet *et al.*, 2004
17β-estradiol (E2)	SW: 0.60 DW: 0.70	Germany	Kuch and Ballschmiter, 2001
	SW: 2–5	Italy	Lagana *et al.*, 2004
	SW: 1.7–2.1	France	Cargouet *et al.*, 2004
	SW: <LOD	USA	Vanderford *et al.*, 2003
17a-ethinylestradiol (EE2)	SW: 0.80 DW: 0.35	Germany	Kuch and Ballschmiter, 2001
	SW: n.d.-1	Italy	Lagana *et al.*, 2004
	SW: 1.3–1.4	France	Cargouet *et al.*, 2004
	SW: 3.6–14	Nevada	Vanderford *et al.*, 2003
Estriol (E3)	SW:2–6	Italy	Lagana *et al.*, 2004
	SW: 1.8–2.2	France	Cargouet *et al.*, 2004

SW: surface water; DW: drinking water; LOD: Limit of Detection.

Table 1.7 Occurrence of steroid hormones in wastewater

Compound	Concentration (ng L^{-1})	Country	Reference
Estrone (E1)	1–100	Canada	Servos *et al.*, 2005
	4–7	France	Cargouet *et al.*, 2004
17β-estradiol (E2)	1–15	Canada	Servos *et al.*, 2005
	5–9	France	Cargouet *et al.*, 2004
17-ethinylestradiol (EE2)	3–5	France	Cargouet *et al.*, 2004

Regarding the fate of steroid hormones in the environment, these compounds are highly hydrophobic, low volatile and present low polarities with octanol–water partition coefficients (K_{ow}) ranging between 10^3 and 10^5. Due to the above, they are significantly sorbed on the suspended solids and sediments. Lai *et al.* (2000) investigated the distribution of several estrogens between water column and sediments and they reported that they can be rapidly removed from water phase to sediments. Additionally, Jürgens *et al.* (1999) indicated that estrogens ended up to bed sediment at percentage up to 92% during the first 24 h.

Biodegradation of estrogens can be occurred in natural environments. Jürgens *et al.* (2002) studied the behaviour of E2 and EE2 in surface waters. The authors concluded that microorganisms capable to transform E2 to E1 were present in river water. As a result, half-lives ranging from 0.2 to 9 days (at 20°C) were calculated. E1 was further degraded at similar rates. On the contrary, EE2 found to be more resistant to biodegradation but susceptible to photodegradation. Interconversion between E1 and E2 occurs, favoring E1 (Birkett and Lester, 2003). This explains the highest E1 concentrations in aquatic environment in many cases.

Furthermore, bioaccumulation of these compounds by several organisms has been reported. Larsson *et al.* (1999) established bioconcentration factors (BCFs) for E1, E2 and EE2 on juvenile rainbow trout. BCF values ranged between 104 and 106. Lai *et al.* (2002) investigated the possible uptake and accumulation of steroid compounds by the alga *Chlorella vulgaris*. They calculated a BCF of 27 for E1, within 48 h under both light and dark conditions. Gomes *et al.* (2004) studied the bioaccumulation of E1 in *Daphnia magna* and found that uptake of the compound via aqueous phase occurred within the first 16 h and a BCF value equal to 228 was estimated. According to the same study, feeding of *Daphnia magna* with algae contaminated with E1, resulted in a partitioning factor of 24 for the crustacean. This fact was an indication of possible biomagnification of E1 via food.

1.4.2 Effects

Steroid hormones are mainly known for their estrogenic action. These compounds present higher endocrine disrupting activity compared with other chemicals (Christiansen *et al.,* 1998). They cause effects such as feminization of male fish and induction of vitellogenesis. Even at concentrations close to detection limits, steroids can cause deleterious effects. For instance, concentrations of 17β-ethinylestradiol (EE2) at level of 0.1 ng L^{-1}, induce the expression of vitellogenin in fish (Purdum *et al.,* 1994). Additionally, affection of sex differentiation (Van Aerle *et al.,* 2002) and fecundity of organisms have been reported. At concentration of 4 ng L^{-1} of EE2, the development of normal secondary sexual characteristics on male fathead minnows was prevented (Länge *et al.,* 2001). Few ng L^{-1} of estradiols also led to induction of vitellogenin in juvenile rainbow trout (Thorpe *et al.,* 2001).

Steroids can cause adverse effects not only in fish but also in other organisms such as amphibians, reptiles or invertebrate. For instance, an oral dosage of EE2 in the range of 0.005–0.09 mg Kg^{-1} d^{-1} can produce carcinogenic effects in female mice (Seibert, 1996). According to Palmer and Palmer (1995), 1 μg g^{-1} of E2 can induce vitellogenesis both in frogs and turtles during a week. Furthermore, inhibition of barnacle settlement due to E2 has been reported (Billinghurst *et al.,* 1998). Effects on plants have also been reported in the literature. Shore *et al.* (1992) indicated that irrigation of alfalfa with water containing estrogens such as E1 and E2, affected the growth of plants.

Humans are also affected by steroids. EE2 has been linked with prostate cancer development (Hess-Wilson and Knudsen, 2006). E2 has also been related with diseases like breast cancer and endometriosis (Dizerega *et al.,* 1980; Thomas, 1984).

1.4.3 Legislation

Control of steroid hormones is a difficult issue due to the fact that natural estrogens cannot be banned or replaced by other compounds. Despite the difficulties, several countries have activated towards this direction. European Union has banned the use of hormones as growth promoters in food-production animals according to the Directive 88/146/EEC. Similarly, the use of compounds such as progesterone, testosterone, estradiol, zeranol and trenbolone acetate for animal food production has been regulated by the US Food and Drug Administration (FDA) and by FAO/WHO.

1.5 SURFACTANTS AND PERSONAL CARE PRODUCTS

Surfactants are a group of synthetic organic compounds consisting of a polar head group and a nonpolar hydrocarbon tail. They are widely used in detergents, textiles, polymer, paper industries and their major classes are anionic (e.g., linear alkylbenzene sulphonates), cationic (e.g., quaternary ammonium compounds) and non ionic surfactants (e.g., alkylphenol ethoxylates) (Ying *et al.,* 2005). Among them, alkylphenol ethoxylates (APEs) constitute a large portion of the surfactant market (production equal to 500.000 t in 1997) (Renner *et al.,* 2007). Significant scientific attention has been given during the last decade to nonylphenol ethoxylates (NPE) which represent almost 80% of the worldwide production of APEs (Brook *et al.,* 2005). The microbial breakdown of NPEs results to the formation of nonylphenol (NP, $C_{15}H_{24}O$) (Figure 1.10) which is much more toxic than the parent compounds and induces estrogenic effects in several aquatic organisms (Birkett and Lester, 2003; Soares *et al.,* 2008).

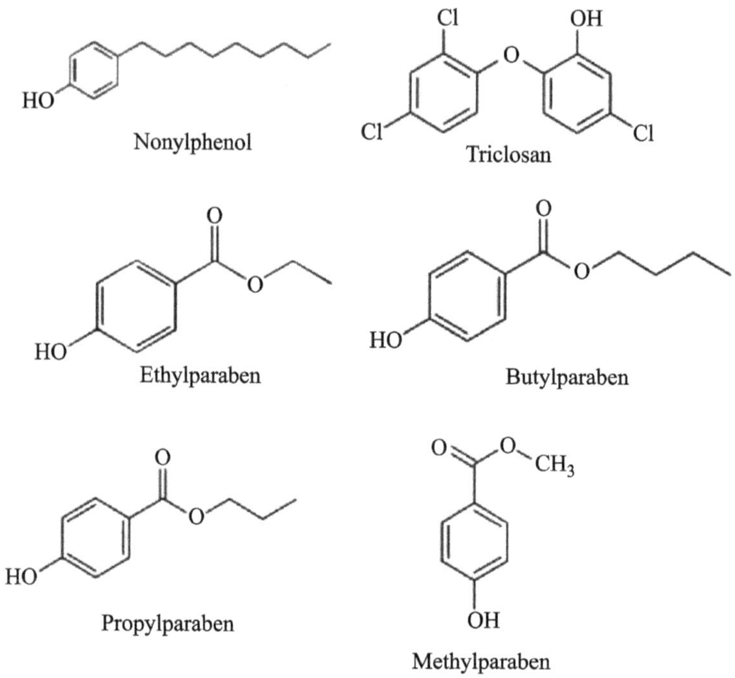

Figure 1.10 Chemical structures of selected surfactants and personal care products

Personal care products include products used for beautification and personal hygiene (skin care products, soaps, shampoos, dental care). These products contain significant concentrations of synthetic organic chemicals such as antimicrobial disinfectants (e.g., triclosan, triclocarban), preservatives (e.g., methylparaben, $C_8H_8O_3$; ethylparaben, $C_9H_{10}O_3$; butylparaben, $C_{11}H_{14}O_3$; propylparaben, $C_{10}H_{12}O_3$) and sunscreen agents (e.g., benzophenone-3, octyl methoxycinnamate) which are introduced to the aquatic environment during regular use (Ternes *et al.,* 2003; Kunz and Fent, 2006). Among them, triclosan (TCS, $C_{12}H_7Cl_3O_2$) and parabens (Figure 1.10) seem to be compounds of significant research and practical interest due to their wide use and toxicological properties (Kolpin *et al.,* 2002). Parabens are used in more than 22000 cosmetic products (Andersen, 2008), while approximately 350 t of TCS are produced annually in Europe for commercial applications (Singer *et al.,* 2002).

1.5.1 Fate

The major source of all these compounds in the environment is the discharge of wastewater. So far, there are several studies investigating their elimination in WWTSs. Regarding NP, contradictory results have been reported for its removal, ranging from minus 9% (Stasinakis *et al.*, 2008) to 98% (Planas *et al.*, 2002; Gonzalez *et al.*, 2007; Jonkers *et al.*, 2009). These differences are due to the fact that NP can be formed during activated sludge process from the biotransformation of NPEs (Ahel *et al.*, 1994). On the other hand, removal efficiency of TCS seem to be more consistent and exceed 90% in most published papers (Heidler and Halden, 2006; Stasinakis *et al.*, 2008). Regarding parabens, in a recent study it was reported that they are almost totally removed during wastewater treatment (Jonkers *et al.*, 2009). The main mechanisms, affecting the removal of these compounds from the dissolved phase of wastewater, are adsorption on the suspended solids and biotransformation to unknown metabolites (Ahel *et al.*, 1994; Heidler and Halden, 2007; Stasinakis *et al.*, 2007; Stasinakis *et al.*, 2008; Stasinakis *et al.*, 2009b). The partial elimination of some of the aforementioned substances (e.g., nonylphenol) during wastewater treatment or/and the disposal of untreated wastewater in the environment result to frequent detection of these compounds in surface waters (Kolpin *et al.*, 2002). Trace concentrations of NP have also been detected in drinking water (Petrovic *et al.*, 2003). In Table 1.8, a few recent data are given concerning the concentrations of these compounds in treated wastewater and surface water.

NP is a hydrophobic compound (log K_{ow} equal to 4.48) with low solubility in water, therefore it partitions mainly to organic matter (John *et al.*, 2000). In natural waters, NP can be photodegraded with a half-life of 10 to 15 hours (Ahel *et al.*, 1994). NP biodegradation is affected by several factors as the existence of aerobic and anaerobic conditions, the type of microorganisms used and their acclimatization on this compound. According to Lalah *et al.* (2003), NP partitions significantly into sediments, while it is resistant to biodegradation in river water and sediment. Other studies have shown that NP can be biodegraded with a slow rate under aerobic conditions in river sediment (half lives ranging between 14 to 99 days) (Yuan *et al.*, 2004) or under anaerobic conditions in mangrove sediments (half lives ranging between 53 to 87 days) (Chang *et al.*, 2009). On the other hand, experiments using river water – sediment and groundwater – aquifer material showed that NP can be rapidly degraded under aerobic and anaerobic conditions (half lives ranging between 0.4 to 1.1 days) due to biotic and abiotic factors (Sarmah and Northcott, 2008). In a recent study, it has been shown that NP biodegradation is differentiated for the different isomers which compose this chemical (Gabriel *et al.*, 2008).

Table 1.8 Concentrations of surfactants and personal care products in water samples

Substance	Concentration (ng L^{-1})	Country	Reference
Surface Water			
NP	<29–195	Switzerland	Jonkers *et al.* (2009)
Methylparaben	3.1–17	Switzerland	Jonkers *et al.* (2009)
Ethylparaben	<LOD-1.6	Switzerland	Jonkers *et al.* (2009)
Propylparaben	<LOD-5.8	Switzerland	Jonkers *et al.* (2009)
Butylparaben	<LOD-2.8	Switzerland	Jonkers *et al.* (2009)
NP	36–33231	China	Peng *et al.* (2008)
Methylparaben	<LOD-1062	China	Peng *et al.* (2008)
Propylparaben	<LOD-2142	China	Peng *et al.* (2008)
TCS	35–1023	China	Peng *et al.* (2008)
NP	0.1–7300	China	Shao *et al.* (2005)
Treated Wastewater			
NP	<LOD-281	Switzerland	Jonkers *et al.* (2009)
Methylparaben	4.6–423	Switzerland	Jonkers *et al.* (2009)
Ethylparaben	<LOD-17	Switzerland	Jonkers *et al.* (2009)
Propylparaben	<LOD-28	Switzerland	Jonkers *et al.* (2009)
Butylparaben	<LOD-12	Switzerland	Jonkers *et al.* (2009)
TCS	<LOD-6880	Greece	Stasinakis *et al.* (2008)
TCS	80–400	Spain	Gomez *et al.* (2007)

LOD: Limit of Detection

TCS is slightly soluble in water, hydrolytically stable and relatively non-volatile (Mc Avoy *et al.*, 2002). Due to the hydrophobicity of its protonated form (log K_{ow} equal to 5.4), it can be sorded to the suspended solids (Singer *et al.*, 2002). TCS is subjected to photolytic transformation in surface waters (Tixier *et al.*, 2002). Photodegradation experiments with natural sunlight showed that its elimination was followed by formation of the more toxic metabolite 2,7/2,8-dibenzodichloro-p-dioxin (Mezcua *et al.*, 2004). In another study, Latch *et al.* (2005) reported that during TCS photodegradation, 2,8-dichlorodibenzo-p-dioxin and 2,4-dichlorophenol are produced. Despite the fact that there are a few data reporting TCS biodegradation in activated sludge process (Federle *et al.*, 2002; Stasinakis *et al.*, 2007) and soil (Ying *et al.*, 2007), there is a lack of data for its biodegradation potential in surface waters.

Regarding parabens, so far there is a lack of data on their fate in the aquatic environment. In a recent study investigating photodegradation and biodegradation of butylparaben, it was reported that this compound is highly stable against sunlight, while it is biodegraded in riverine water (Yamamoto *et al.*, 2007).

1.5.2 Effects

NP is considered as an endocrine disruptor (Birkett and Lester, 2003). A great number of data are available in the literature regarding its effects on aquatic organisms, while several review papers have been published (Staples *et al.*, 2004; Vazquez – Duhalt *et al.*, 2005; Soares *et al.*, 2008). These papers indicate that the effects of NP are very diverse and they are depended on the test organism (species, stage of development) and the characteristics of the environment. In a recent study, Hirano *et al.* (2009) reported that environmentally relevant concentrations of NP can disrupt growth of the mysid crustacean *Americamysis bahia*. Moreover, exposure of Atlantic salmon smolts to 10 μg L^{-1} NP for 21 days caused direct and delayed mortalities (Lerner *et al.*, 2007). In another study, Schubert *et al.* (2008) reported that a mixture of NP, E1 and E2 at concentration levels of a few ng L^{-1} would not adversely affect reproductive capability of brown trout. Due to the fact that NP is composed of several isomers, recent studies have been focused on correlating isomer molecular structure and endocrine activity (Gabriel *et al.*, 2008; Preuss *et al.*, 2009). Bioaccumulation of NP has been observed in algae, fish and aquatic birds (Ahel *et al.*, 1993; Hu *et al.*, 2007).

TCS toxicity has been investigated in algae, invertebrates, and fish (Orvos *et al.*, 2002; Dussault *et al.*, 2008). Recent studies have also shown that this compound may act as an endocrine disruptor (Veldhoen *et al.*, 2006; Kumar *et al.*, 2009). According to Veldhoen *et al.* (2006), environmentally relevant TCS concentrations altered thyroid hormone receptor mRNA expression in the American bullfrog. TCS is rapidly accumulated in algae and freshwater snails (Coogan and La Point, 2008). Moreover, it has been reported the bioaccumulation of TCS and its biotransformation product, methyl-TCS, in fish (Balmer *et al.*, 2004). Recently, TCS was detected in the plasma of dolphins at concentrations ranging between 0.025 to 0.27 ng g^{-1} wet weight, indicating its bioaccumulation in marine mammals (Fair *et al.*, 2009).

The presence of parabens in human urine and their ability to penetrate human skin have been demonstrated by several authors (Ye *et al.*, 2006; El Hussein *et al.*, 2007; Darbre and Harvey, 2008). These compounds seem to have low acute toxicity (Andersen, 2008), while they show little tendency to bioaccumulate in aquatic organisms (Alslev *et al.,* 2005). Recently, suspicions have been raised concerning their potential for causing endocrine disrupting effects in rainbow trout (propyl and butyl parabens) (Bjerregaard *et al.,* 2003; Alslev *et al.,* 2005) and medaka (propyl paraben) (Inui *et al.,* 2003).

1.5.3 Legislation

Due to the fact that surfactants and personal care products are emerging contaminants, there are limited regulations for their concentrations in the environment. Among the studied compounds, regulations have been established only for nonylphenols. Specifically, NP and its ethoxylates have been listed as priority substances in the Water Framework Directive (EU, 2001) and most of their uses are currently regulated (EU, 2003). EPA prepared a guideline setting quality criteria for NP in freshwater and saltwater (Brooke and Thursby, 2005). Moreover, Canada has set stringent water quality guidelines for NP and NPEs (Enironment Canada, 2001; 2002). European Union in an attempt to set some limit values for trace organic contaminants in sludge, proposed in a Working Document a limit value of 50 $\mu g\ g^{-1}$ dry weight for NPEs (sum of NP, NP1EO, NP2EO) (EU, 2000). However, at the moment, only few countries such as Switzerland and Denmark have legislation about the concentrations of NPEs in sewage sludge (JRC, 2001).

1.6 PERFLUORINATED COMPOUNDS

Perfluorinated compounds (PFCs) have been produced since 1950. They are used in a great number of industrial and consumer applications due to their physico-chemical characteristics such as thermal and chemical stability, surface active properties and low surface free energy (Lehmler, 2005). The bond C–F is particularly strong and as a result these compounds are resistant to various modes of degradation, such as oxidation, reduction and reaction with acids and bases (Kissa, 2001). Among PFCs, the most commonly studied substances are perfluorinated sulfonates (PFAS) and perfluorinated carboxylates (PFCA). These molecules consist of one perfluorinated carbon chain and one sulfonic (PFAS) or carboxylic group (PFCA). Among these, several compounds have been detected in the environment such as perfluorononanoic acid (PFNA, $C_9HF_{17}O_2$), perfluorodecanoic acid (PFDA, $C_{10}HF_{19}O_2$), perfluoroundecanoic acid (PFUnA, $C_{11}HF_{21}O_2$), perfluorododecanoic acid (PFDoA, $C_{12}HF_{23}O_2$) and perfluorooctanesulfonamide (PFOSA, $C_8H_2F_{17}NO_2S$).

However, perfluorooctane sulfonate (PFOS, $C_8HF_{17}O_3S$) and perfluorooctanoate (PFOA, $C_8HF_{15}O_2$) seem to be of greatest concern (Figure 1.11), due to their extended uses, the concentration levels detected, their behavior and toxicity. The production of PFOS was almost 3500 metric tons in 2000 (Lau *et al.,* 2007). Due to the fact that 3M Company, the major manufacturer of PFOS, phased out production in 2002, the global production of this chemical decreased to 175 metric tons by 2003 (3M Company, 2003). Regarding PFOA, its production was estimated to be almost 500 metric tons in 2000, while it was increased to 1200 metric tons by 2004 (Lau *et al.,* 2007). PFOS and its precursors are used in many applications such as food packing materials, surfactants in diverse cleaning agents, cosmetics, fire-fighting foams, electronic and photographic devices, (OECD, 2002 OECD (Organization for Economic Co-operation and Development), 2002. Co-operation on existing chemicals. Hazard assessment of perfluorooctane sulfonate (PFOS) and its salts. ENV/JM/RD(2002)17/FINAL, Paris.Kissa 2001). PFOA is mainly used during the production of certain fluoropolymers such as polytetrafluoroethylene (PTFE) and to a lesser extent in other industrial applications (OECD, 2005).

$CF_3(CF_2)_nCOO^-$

Perfluorocarboxylate (PFCA)

$CF_3(CF_2)_nSO_3^-$

Perfluoroalkyl sulfonate (PFAS)

Figure 1.11 Chemical structures of PFCA and PFAS

1.6.1 Fate

PFCs are commercially synthesized by electrochemical fluorination or telomerization (Fromme *et al.,* 2009). However, PFCs can also be formed in the environment from biotic and abiotic transformation of commercially synthesized precursors. For instance, it has been reported that perfluorooctane sulfonamides can be biotransformed to PFOS (Tomy *et al.,* 2004), while fluorotelomer alcohols (FTOH) can be subsequently transformed into PFOA (Wang *et al.,* 2005).

PFCs have been detected in potable water, surface water, groundwater and wastewater worldwide (Table 1.9). Moreover, they have been detected in remote areas, reflecting the widespread global pollution for these compounds (Giesy and Kannan, 2001; Houde *et al.,* 2006). So far, a few data have been reported for the fate of PFOA and PFOS in the environment and the mechanisms affecting their distribution and transport. PFOA and PFOS are considered stable compounds due to their resistance in abiotic and biotic degradation (Giesy and Kannan, 2002). Yamashita *et al.* (2005) reported that there is a long range transport of these compounds by oceanic currents, while other authors proposed the atmospheric transport of volatile precursor chemicals and their transformation to PFOS and PFOA as an explanation for the presence of these anthropogenic chemicals in remote regions (Young *et al.,* 2007).

In a recent study investigating the occurrence of organic micropollutants in European rivers, PFOA and PFOS were detected in 97% and 94% of samples, respectively (Loos *et al.,* 2009). The discharge of municipal wastewater seems to be one of the major routes introducing these compounds into the aquatic environment (Sinclair and Kannan, 2006; Becker *et al.,* 2008). In some papers, the concentrations of these compounds in effluent wastewater are higher that those detected in influent wastewater, indicating possible biodegradation of their precursors during biological treatment processes (Sinclair and Kannan, 2006; Murakami *et al.,* 2009).

1.6.2 Effects

The toxicity of PFOS and PFOA has been studied extensively. Hepatotoxicity, immunotoxicity, developmental toxicity, hormonal effects and a carcinogenic potency are the effects of main concern (Lau *et al.,* 2004; Lau *et al.,* 2007). Animal studies have shown that these compounds are mainly distributed to serum, liver and kidney (Seacat *et al.,* 2002; Hundley *et al.,* 2006). The elimination half-lives of PFOA and PFOS are significantly differentiated for different species or different gender of the same species, ranging from few hours to 30 days (Kemper, 2003; Butenhoff *et al.,* 2004).

PFCs have also been detected in human blood and tissue samples from occupationally and non-occupationally exposed humans throughout the world (Kannan *et al.,* 2004; Calafat *et al.,* 2006; Olsen *et al.,* 2007). Food intake and drinking water consumption seems to be the major contemporary exposure pathway for the background population (Vestergren and Cousins, 2009). According to a recent study, PFOA concentrations in human blood range between 2 and 8 $\mu g\ L^{-1}$ for background exposed population in industrialized countries (Vestergren and Cousins, 2009). Epidemiologic data related to PFC exposure are limited and they have been mainly performed on PFC production plant workers. In general, serum fluorochemical levels have not been associated with adverse health effects (Lau *et al.*, 2007). Studies in retirees from PFC production facilities showed a mean elimination half-life of 3.8 years and 5.4 years for PFOA and PFOS, respectively (Olsen *et al.*, 2007). In a recent review paper, Fromme *et al.* (2009) describe the different pathways which are responsible for human exposure to PFCs (exposure via inhalation from outdoor and indoor air, oral exposure via food and water consumption). Regarding the acute toxicity of these compounds in aquatic organisms, it seems that at concentration levels which are similar to those detected in the environment, no acute toxicity of PFOS or PFOA has been observed (Sanderson *et al.*, 2004; Li, 2009).

The persistence of PFCs in the environment and their potential to accumulate and biomagnificate in the food chain is a matter of significant toxicological concern. PFCs have been detected in serum or plasma of wildlife worldwide (Keller *et al.*, 2005; Tao *et al.*, 2006). Determination of PFOS concentrations in water and organisms at various trophic levels (Kannan *et al.*, 2005) showed that its concentrations in benthic invertebrates were 1000-fold greater than those in surrounding water. Moreover, a biomagnification factor (BMF) equal to 20 was calculated for bald eagles (Kannan *et al.*, 2005). Bioconcentration experiments with different PFCs (Martin *et al.*, 2003) showed that bioconcentration factors (BCFs) increased with increasing length of the perfluoroalkyl chain, while carboxylates had lower BFCs than sulfonates of equal perfluoroalkyl chain length. As a result, BCF values equal to 27 ± 9.7, 4300 ± 570 and 40000 ± 4500 (L/Kg) were calculated in blood of rainbow trout for PFOA, PFOS and PFDoA, respectively (Martin *et al.*, 2003).

1.6.3 Legislation

From a regulatory point of view, PFOS has been classified as very persistent, very bioaccumulative and toxic compound, fulfilling the criteria for being considered as a persistent organic pollutant under the Stockholm Convention (EU, 2006). In May 2009, PFOS was added to Annex B of Stockholm Convention on Persistent Organic Pollutants. Canada banned the importation and use of several long chain perfluorinated carboxylic acids due to their effects on human and environment (Canadian Government Department of the Environment, 2008). In 2002, EPA issued a Significant New Use Rule regulating the import and production of several perfluorooctanyl-based chemicals (EPA, 2002). Finally, in 2006, EPA established the 2010/15 PFOA Stewardship Program. Targets of this Program are the reduction of global facility emissions and product content of PFOA and related chemicals by 95 percent up to 2010 and the elimination of emissions and product content of these compounds up to 2015.

Table 1.9 Concentrations of PFCs in water samples

Substance	Concentration (ng L^{-1})	Country	Reference
Surface Water			
PFOA	0.6–15.9	Italy	Loos *et al.*, 2007
PFOS	<LOD-38.5	Italy	Loos *et al.*, 2007
PFNA	0.2–16.2	Italy	Loos *et al.*, 2007
PFDA	<LOD-10.8	Italy	Loos *et al.*, 2007
PFUnA	0.1–38.0	Italy	Loos *et al.*, 2007
PFDoA	<LOD-14.1	Italy	Loos *et al.*, 2007
PFOS	9.3–56	USA	Plumlee *et al.*, 2008
PFOA	<LOD-36	USA	Plumlee *et al.*, 2008
PFDA	<LOD-19	USA	Plumlee *et al.*, 2008
PFOA	0.8–14	Germany	Becker *et al.*, 2008
PFOS	<LOD-15	Germany	Becker *et al.*, 2008
Groundwater			
PFOS	19–192	USA	Plumlee *et al.*, 2008
PFOA	<LOD-28	USA	Plumlee *et al.*, 2008
PFDA	<LOD-19	USA	Plumlee *et al.*, 2008
Potable Water			
PFOA	1.0–2.9	Italy	Loos *et al.*, 2007
PFOS	6.2–9.7	Italy	Loos *et al.*, 2007
PFNA	0.3–0.7	Italy	Loos *et al.*, 2007
PFDA	0.1–0.3	Italy	Loos *et al.*, 2007
PFUnA	0.1–0.4	Italy	Loos *et al.*, 2007
PFDoA	0.1–2.8	Italy	Loos *et al.*, 2007
PFOS	<LOD-22	Japan	Takagi *et al.*, 2008
PFOA	2.3–84	Japan	Takagi *et al.*, 2008
Treated Municipal Wastewater			
PFOS	3–68	USA	Sinclair and Kannan, 2006
PFOA	58–1050	USA	Sinclair and Kannan, 2006
PFNA	<LOD-376	USA	Sinclair and Kannan, 2006
PFDA	<LOD-47	USA	Sinclair and Kannan, 2006
PFOS	20–190	USA	Plumlee *et al.*, 2008
PFOA	12–190	USA	Plumlee *et al.*, 2008

(*continued*)

Table 1.9 (*continued*)

Substance	Concentration (ng L^{-1})	Country	Reference
PFDA	<LOD-11	USA	Plumlee *et al.*, 2008
PFNA	<LOD-32	USA	Plumlee *et al.*, 2008
PFOA	8.7–250	Germany	Becker *et al.*, 2008
PFOS	2.4–195	Germany	Becker *et al.*, 2008

LOD: Limit of Detection

1.7 REFERENCES

3M Company (2003) Environmental and health assessment of perfluorooctanesulfonate and its salts. US EPA Administrative Record. AR-226–1486.

Ahel, M., Giger, W. and Koch, M. (1994) Behaviour of alkylphenol polyethoxylate surfactants in the aquatic environment – I. Occurrence and transformation in sewage treatment. *Water Res.* **28**, 1131–1142.

Ahel, M., Mc Evoy, J. and Giger, W. (1993) Bioaccumulation of the lipophilic metabolites of nonionic surfactants in freshwater organisms. *Environ. Pollut.* **79**, 243–248.

Ahel, M., Scully, F. E., Hoigne, J. and Giger, W. (1994) Photochemical degradation of nonylphenol and nonylphenol polyethoxylates in natural-waters. *Chemosphere* **28**, 1361–1368.

Albanis, T. A., Hela, D. G., Lambropoulou, D. A. and Sakkas, V. A. (2004) Gas chromatographicemass spectrometric methodology using solid phase microextraction for the multiresidue determination of pesticides in surface waters (N.W. Greece). *Int. J. Environ. An. Ch.* **84**, 1079–1092.

Alslev, B., Korsgaard, B. and Bjerregaard, P. (2005) Estrogenicity of butylparaben in rainbow trout Oncorhynchus mykiss exposed via food and water. *Aquat. Toxicol.* **72**, 295–304.

Andersen, F. A. (2008) Final amended report on the safety assessment of methylparaben, ethylparaben, propylparaben, isopropylparaben, butylparaben, isobutylparaben, and benzylparaben as used in cosmetic products. *Int. J. Toxicol.* **27**, 1–82.

Andreozzi, R., Raffaele, M. and Paxéus, N. (2003) Pharmaceuticals in STP effluents and their solar photodegradation in aquatic environment. *Chemosphere* **50**, 1319–1330.

Araújo, T. M., Campos, M. N. N. and Canela, M. C. (2007) Studying the photochemical fate of methyl parathion in natural waters under tropical conditions. *Int. J. Environ. An. Ch.* **87**, 937–947.

Ashton, D., Hilton, M. and Thomas, K. V. (2004) Investigating the environmental transport of human pharmaceuticals to streams in the United Kingdom. *Sci. Total Environ.* **333**, 167–184.

Balmer, M. E., Poiger, T., Droz, C., Romanin, K., Bergqvist, P. A. and Mueller J. F. (2004) Occurrence of methyl Triclosan, a transforation product of the bactericide triclosan, in fish from various lakes in Switzerland. *Environ Sci. Technol.* **38**, 390–395.

Becker, A. M., Gerstmann, S. and Frank, H. (2008) Perfluoroctane surfactants in waste waters, the major source of river pollution. *Chemosphere* **72**, 115–121.

Bennoti, M. J. and Brownawell, B. J. (2009) Microbial degradation of pharmaceuticals in estuarine and coastal seawater. *Environ. Pollut.* **157**, 994–1002.

Billinghurst, Z., Clare, A. S., Fileman, T., McEvoy, J., Readman, J. and Depledge, M. H. (1998) Inhibition of barnacle settlement by the environmental oestrogen 4-nonylphenol and the natural oestrogen 17 beta oestradiol. *Mar. Pollut. Bull.* **36**, 833–839.

Birkett, J. W. and Lester, J. N. (2003) Endocrine disrupters in wastewater and sludge treatment processes. CRC Press LLC, Florida.

Bjerregaard, P., Andersen, D. N., Pedersen, K. L., Pedersen, S. N. and Korsgaard, B. (2003) Estrogenic effect of propylparaben (propylhydroxybenzoate) in rainbow trout Oncorhynchus mykiss after exposure via food and water. *Comp. Biochem. Phys. Part C.* **136**, 309–317.

Blackwell, P. A., Kay, P. and Boxall, A. B. A. (2007) The dissipation and transport of veterinary antibiotics in a sandy loam soil. *Chemosphere* **67**, 292–299.

Blasiak, J., Kleinwachter, V., Walter, Z. and Zaludova, R. (1995) Interaction of organophosphorus insecticide methyl parathion with calf thymus DNA and a synthetic DNA duplex. *Z Naturforsch C.* **50**, 820–823.

Botella, B., Crespo, J., Rivas, A., Cerrillo, S., Olea-Serrano, M.-F. and Olea, N. (2004) Exposure of women to organochlorine pesticides in Southern Spain. *Environ. Res.* **96**, 34–40.

Bound, J. P., Kitsou, K. and Voulvoulis, N. (2006) Household disposal of pharmaceuticals and perception of risk to the environment. *Environ. Toxicol. Phar.* **21**, 301–307.

Boxall, A. B., Kolpin, D. W., Halling-Sorensen, B. and Tolls, J. (2003) Are veterinary medicines causing environmental risks? *Environ. Sci. Technol.* **37**, 286–294.

Brook, D., Crookes, M., Johnson, I., Mitchell, R. and Watts, C. (2005) National Centre for Ecotoxicology and Hazardous Substances, Environmental Agency, Bristol U.K. 2005. Prioritasation of alkylphenols for environmental risk assessment.

Brooke, L. and Thursby, G. (2005) Ambient aquatic life water quality criteria for nonylphenol. Washington DC, USA: Report for the United States EPA, Office of Water, Office of Science and Technology.

Brown J. N., Paxeus N. and Forlin L. (2007) Variations in bioconcentration of human pharmaceuticals from sewage effluents into fish blood plasma. *Environ. Toxicol. Phar.* **24**, 267–274.

Buser, H.-R., Poiger, T. and Müller, M. D. (1998) Occurrence and fate of the pharmaceutical drug diclofenac in surface waters: rapid photodegradation in a lake. *Environ. Sci. Technol.* **33**, 3449–3456.

Butenhoff, J. L., Kennedy, G. L., Hindliter, P. M., Lieder, P. H., Hansen, K. J., Gorman, G. S., Noker, P. E. and Thomford, P. J. (2004) Pharmacokinetics of perfluorooctanoate in Cynomolgus monkeys. *Toxicol. Sci.* **82**, 394–406.

Calafat, A. M., Kuklenyik, Z., Caudill, S. P., Reidy, J. A. and Needham, L. L. (2006) Perfluorochemicals in pooled serum samples from the United States residents in 2001 and 2002. *Environ. Sci. Technol.* **40**, 2128–2134.

Canadian Government Department of the Environment (2008). Perfluorooctanesulfonate and its salts and certain other compounds regulations. Canada Gazette, Part II **142**, 322–1325.

Carafa, R., Wollgast, J., Canuti, E., Ligthart, J., Dueri, S., Hanke, G., Eisenreich, S. J., Viaroli, P. and Zaldívar, J. M. (2007) Seasonal variations of selected herbicides and related metabolites in water, sediment,seaweed and clams in the Sacca di Goro

coastal lagoon (Northern Adriatic). *Chemosphere* **69**, 1625–1637.
Cargouet, M., Perdiz, D., Mouatassim-Souali, A., Tamisier-Karolak, S. and Levi, Y. (2004) Assessment of River Contamination by Estrogenic Compounds in Paris Area (France). *Sci. Total Environ.* **324**, 55–66.
Castillo, M., Domingues, R., Alpendurada, M. F. and Barceló, D. (1997) Persistence of selected pesticides and their phenolic transformation products in natural waters using off-line liquid solid extraction followed by liquid chromatographic techniques. *Anal. Chim. Acta* **353**, 133–142.
Chang, B. V., Lu, Z. J. and Yuan, S. Y. (2009) Anaerobic degradation of nonylphenol in subtropical mangrove sediments. *J. Hazard. Mater.* **165**, 162–167.
Chen, C. W., Hurd, C., Vorojeikina, D. P., Arnold, S. F. and Notides, A. C. (1997) Trascriptional activation of the human estrogen receptor by DDT isomers and metabolites in yeast and MCF-7 cells. *Biochem. Pharmacol.* **53**, 1161–1172.
Choi, K., Kim, Y., Jung, J., Kim, M. H., Kim, C. S., Kim, N. H. and Park, J. (2008) Occurrences and ecological risk of roxithromycin, trimethoprim and chloramphenicol in the Han river, Korea. *Environ. Toxicol. Chem.* **27**, 711–719.
Christensen, A. M., Ingerslev, F. and Baun, A. (2006) Ecotoxicity of mixtures of antibiotics used in aquacultures. *Environ. Toxicol. Chem.* **25**, 2208–2215.
Christiansen, T., Korsgaard, B. and Jespersen, A. (1998) Effects of nonylphenol and 17β-oestradiol on vitellogenin synthesis, testicular structure and cytology in male eelpout Zoarces viviparous. *J. Exp. Biol.* **201**, 179–192.
Claver, A., Ormad, P., Rodríguez, L. and Ovelleiro, J.-L. (2006) Study of the presence of pesticides in surface waters in the Ebro river basin (Spain). *Chemosphere* **64**, 1437–1443.
Cleuvers, M. (2004) Mixture toxicity of the anti-inflammatory drugs diclofenac, ibuprofen, naproxen, and acetylsalicylic acid. *Ecotox. Environ. Safe.* **59**, 309–315.
Coogan, M. A. and La Point, T. W. (2008) Snail bioaccumulation of triclocarban, triclosan, and methyltriclosan in a North Texas, USA, stream affected by wastewater treatment plant runoff. *Environ. Toxicol. Chem.* **27**, 1788–1793.
Corbet, J. R. (1974) The Biochemical mode of action of pesticides, Academic Press, London.
Costanzo, S. D., Murby, J. and Bates, J. (2005) Ecosystem response to antibiotics entering the aquatic environment. *Mar. Pollut. Bull.* **51**, 218–223.
D'Ascenzo, G., Corcia, A. D., Mancini, A. G. R., Mastropasqua, R., Nazzari, M. and Samperi, R. (2003) Fate of natural estrogen conjugates in municipal sewage transport and treatment facilities. *Sci. Total Environ.* **302**, 199–209.
Dahmardeh-Behrooz, R., Esmaili Sari, A., Bahramifar, N. and Ghasempouri, S. M. (2009) Organochlorine pesticide and polychlorinated biphenyl residues in human milk from the Southern Coast of Caspian Sea, Iran. *Chemosphere* **74**, 931–937.
Darbre, P. D. and Harvey, P. W. (2008) Paraben esters: Review of recent studies of endocrine toxicity, absorption, esterase and human exposure, and discussion of potential human health risks. *J. Appl. Toxicol.* **28**, 561–578.
Dizerega, G. S., Barber, D. L. and Hodgen, G. D. (1980) Endometriosis: role of ovarian steroids in initiation, maintenance, and suppression. *Fertil. Steril.* **33**, 649–653.
Drillia, P., Stamatelatou, K. and Lyberatos, G. (2005) Fate and mobility of pharmaceuticals in solid matrices. *Chemosphere* **60**, 1034–1044.
Dussault, E. B., Balakrishnan, V. K., Sverko, E., Solomon, K. R. and Sibley, P. K. (2008) Toxicity of human pharmaceuticals and personal care products to benthic invertebrates. *Environ. Toxicol. Chem.* **27**, 425–432.
Dzyadevych, S. V., Soldatkin, A. P. and Chovelon, J.-M. (2002) Assessment of the

toxicity of methyl parathion and its photodegradation products in water samples using conductometric enzyme biosensors. *Anal. Chim. Acta* **459**, 33–41.

El Hussein, S., Muret, P., Berard, M., Makki, S. and Humbert, P. (2007) Assessment of principal parabens used in cosmetics after their passage through human epidermis-dermis layer (ex-vivo study). *Exp. Dermatol.* **16**, 830–836.

EMEA (1998) Note for guidance: environmental risk assessment for veterinary medicinal products other than GMO-containing and immunological products, EMEA, London (EMEA/CVMP/055/96).

Environment Canada (2001). Nonylphenol and its ethoxylates: Priority substances list assessment report. Report no. EN40-215-/57E.

Environment Canada (2002). Canadian environmental quality guidelines for nonylphenol and its ethoxylates (water, sediment and soil). Scientific Supporting Document. Ecosystem Helath: Sciencebased solutions report No 1–3. National Guidelines and Standard Office, Environmental Quality Branch, Environment Canada, Ottawa.

EPA (2002). Rules and regulations. United States Federal Register 67, pp. 72854–72867.

EU (1988) Council Directive 88/146/EEC for prohibiting the use in livestock farming of certain substances having a hormonal action.

EU (1998) Council Directive on the Quality of Water Intended for Human Consumption, 98/83/CE.

EU (2000) Working document on sludge, Third Draft, European Union, Brussels, Belgium, April 27, 2000.

EU (2001) European Union, Decision No 2455/2001/EC of the European Parliament and of the council of 20 November 2001 establishing the list of priority substances in the field of water policy and amending directive 2000/60/EC, *Off. J.* L331 (15/12/2001).

EU (2003) Directive 2003/53/EC, Amending for the 26th time the Council directive 76/769/EEC relating to restrictions on the marketing and use of certain dangerous substances and preparations (nonylphenol, nonylphenol ethoxylate and cement), Luxembourg, Luxembourg: European Parliament and the Council of the European Union.

EU (2006) Directive 2006/122/ECOF of the European Parliament and of the Council of 12 December 2006. Official Journal of the European Union, L/372/32–34, 27.12.2006.

EU (European Union), 2004. Directive 2004/27/EC, Amending directive 2001/83/EC on the community code relating to medicinal products for human use. Official Journal of the European Union, L/136/34–57, 30.04.2004.

EU (European Union), 2004. Directive 2004/28/EC, Amending directive 2001/82/EC on the community code relating to veterinary medicinal products. Official Journal of the European Union, L/136/58–84, 30.04.2004.

Fair, P. A., Lee, H. B., Adams, J., Darling, C., Pacepavicius, G., Alaee, M., Bossart, G. D., Henry, N. and Muir, D. (2009) Occurrence of triclosan in plasma of wild Atlantic bottlenose dolphins (Tursiops truncatus) and in their environment. *Environ. Pollut.* **157**, 2248–2254.

Fan, W., Yanase, T., Morinaga, H., Gondo, S., Okabe, T., Nomura, M., Hayes, T. B., Takayanagi, R. and Nawata, H. (2007) Herbicide atrazine activates SF-1 by direct affinity and concomitant co-activators recruitments to induce aromatase expression via promoter II. *Biochem. Bioph. Res. Co.* **355**, 1012–1018.

Farre, M., Perez, S., Kantiani, L. and Barcelo, D. (2008) Fate and toxicity of emerging pollutants, their metabolites and transformation products. *TrAC Trend Anal. Chem.*, **27**, 991–1007.

FDA-CDER (1998) Guidance for Industry-Environmental Assessment of Human Drugs and Biologics Applications, Revision 1, FDA Center for Drug Evaluation and Research, Rockville.

Federle, T. W., Kaiser, S. K. and Nuck, B. A. (2002) Fate and effects of triclosan in activated sludge. *Environ. Toxicol. Chem.* **21**, 1330–1337.

Fent, K., Weston, A. A. and Caminada, D. (2006) Ecotoxicology of human pharmaceuticals. *Aquat. Toxicol.* **76**, 122–159.

Fernadez-Alba, A. R., Hernando, M. D., Piedra, L. and Chisti, Y. (2002) Toxicity evaluation of single and mixed antifouling biocides measured with acute toxicity bioassays. *Anal. Chim. Acta.* **456**, 303–312.

Figueroa, R. A., Leonard, A. and Mackay, A. N. (2004) Modeling tetracycline antibiotics sorption to clays. *Environ. Sci. Technol.* **38**, 476–483.

Fromme, H., Tittlemier, S. A., Volkel, W., Wilhelm, M. and Twardella, D. (2009) Perfluorinated compounds – Exposure assessment for the general population in western countries. *Int. J. Hyg. Envir. Heal.*, **212**, 239–270.

Gabriel, F. L. P., Routledge, E. J., Heidlberger, A., Rentsch, D., Guenther, K., Giger, W., Sumpter, J. P. and Kohler, H. P. E. (2008) Isomer-specific degradation and endocrine disrupting activity of nonylphenols. *Environ. Sci. Technol.* **42**, 6399–6408.

Gangwar, S. K. and Rafiquee, M. Z. A. (2007) Kinetics of the acid hydrolysis of isoproturon in the absence and presence of sodium lauryl sulfate micelles. *Colloid Polym. Sci.* **285**, 587–592.

Gao, J., Liu, L., Liu, X., Zhou, H., Lu, J., Huang, S. and Wang, Z. (2009) The occurrence and spatial distribution of organophosphorous pesticides in Chinese surface water. *B. Environ. Contam. Tox.* **82**, 223–229.

Gatidou, G., Kotrikla, A., Thomaidis, N. S. and Lekkas, T. D. (2004) Determination of the antifouling booster biocides irgarol 1051 and diuron and their metabolites in seawater by high performance liquid chromatography–diode array detector. *Anal. Chim. Acta* **528**, 89–99.

Gatidou, G. and Thomaidis, N. S. (2007) Evaluation of single and joint toxic effects of two antifouling biocides, their main metabolites and copper using phytoplankton bioassays. *Aquat. Toxicol.* **85**, 184–191.

Gatidou, G., Thomaidis, N. S. and Zhou, J. L. (2007) Fate of Irgarol 1051, diuron and their main metabolites in two UK marine systems after restrictions in antifouling paints. *Environ. Int.* **33**, 70–77.

Giacomazzi, S. and Cochet, N. (2004) Environmental impact of diuron transformation: a review. *Chemosphere* **56**, 1021–1032.

Giesy, J. P. and Kannan, K. (2001) Global distribution of perfluoroctane sulfonate in wildlife. *Environ. Sci. Technol.* **35**, 1339–1342.

Giesy, J. P. and Kannan, K. (2002) Perfluorochemicals in the environment. *Environ. Sci. Technol.* **36**, 147–152.

Golfinopoulos, S.K, Nikolaou, A. D., Kostopoulou, M. N., Xilourgidis, N. K., Vagi, M. C. and Lekkas, D. T. (2003) Organochlorine pesticides in the surface waters of Northern Greece. *Chemosphere* **50**, 507–516.

Gomes, R. L., Deacon, H. E., Lai, K. M., Birkett, J. W., Scrimshaw, M. D. and Lester, J. N. (2004) An assessment of the bioaccumulation of estrone in daphnia magna. *Environ. Toxicol. Chem.* **23**, 105–108.

Gómez, M. J., Martínez Bueno, M. J., Lacorte, S., Fernández-Alba, A. R. and Agüera, A. (2007) Pilot survey monitoring pharmaceuticals and related compounds in a sewage treatment plant located on the Mediterranean coast. *Chemosphere* **66**, 993–1002.

Gonzalez, S., Petrovic, M. and Barcelo, D. (2007) Removal of a broad range of surfactants from municipal wastewater – Comparison between membrane bioreactor and conventional activated sludge treatment. *Chemosphere* **67**, 335–343.

Gooddy, D. C., Chilton, P. J. and Harrison, I. (2002) A field study to assess the degradation and transport of diuron and its metabolites in a calcareous soil. *Sci. Total Environ.* **297**, 67–83.

Götz, R., Bauer, O. H., Friesel, P. and Roch, K. (1998) Organic trace compounds in the water of the River Elbe near Hamburg part II. *Chemosphere* **36**, 2103–2118.

Government of Canada, 2006. Perfluorooctane sulfonate and its salts and certain other compounds regulations. *Can. Gazette Part. I* **140**, 4265–4284.

Groning, J., Held, C., Garten, C., Claussnitzer, U., Kaschabek, S. R. and Schlomann, M. (2007) Transformation of diclofenac by the indigenous microflora of river sediments and identification of a major intermediate. *Chemosphere* **69**, 509–516.

Halling-Sørensen, B., Nors Nielsen, S., Lanzky, P. F., Ingerslev, F., Holten Lützhoft, H. C. and Jørgensen, S. E. (1998) Occurrence, fate and effects of pharmaceutical substances in the environment – a review. *Chemosphere* **36**, 357–393.

Heberer, T. (2002a) Occurrence, fate, and removal of pharmaceutical residues in the aquatic environment: a review of recent research data. *Toxicol. Lett.* **131**, 5–17.

Heberer, T. (2002b) Tracking persistent pharmaceutical residues from municipal sewage to drinking water. *J. Hydrol.* **266**, 175–189.

Heidler, J. and Halden, R. U. (2007) Mass balance of triclosan removal during conventional sewage treatment. *Chemosphere* **66**, 362–369.

Hess-Wilson, J. K. and Knudsen, K. E. (2006) Endocrine disrupting compounds and prostate cancer. *Cancer Lett.* **241**, 1–12.

Hignite, C. and Azarnoff, D. L. (1977) Drugs and drug metabolites as environmental contaminants: Chlorophenoxyisobutyrate and salicylic acid in sewage water effluent. *Life Sci.* **20**, 337–341.

Hirano, M., Ishibashi, H., Kim, J. W., Matsumura, N. and Arizono, K. (2009) Effects of environmentally relevant concentrations of nonylphenol on growth and 20-hydroxyecdysone levels in mysid crustacean, Americamysis bahia. *Comp. Biochem. Phys. Part C: Toxicol. Pharm.* **149**, 368–373.

Hirsch, R., Ternes, T. A., Haberer, K. and Kratz, K. L. (1996) Determination of betablockers and β-sympathomimetics in the aquatic environment. *Vom Wasser* **87**, 263–274.

Houde, M., Martin, J. W., Letcher, R. J., Solomon, K. R. and Muir, D. C. G. (2006) Biological monitoring of polyfluoroalkyl substances: A review. *Environ. Sci. Technol.* **40**, 3463–3473.

Hu, J., Jin, F., Wan, Y., Yang, M., An, L., An, W. and Tao, S. (2005) Trophodynamic behavior of 4-nonylphenol and nonylphenol polyethoxylate in a marine aquatic food web from Bohai Bay, North China: Comparison to DDTs. Environ. Sci. Technol. **39**, 4801–4807.

Humburg, N. E., Colby, S. R. and Hill, E. R. (1989) Herbicide Handbook of the Weed Science Society of America (sixth ed.), Weed Science Society of America, Champaign, IL.

Hundley, S. G., Sarrif, A. M. and Kennedy, G. L. (2006) Absorption, distribution, and excretion of ammonium perfluorooctanoate (APFO) after oral administration to various species. *Drug Chem. Toxicol.* **29**, 137–145.

IARC/WHO (1991) Occupational exposures in insecticide application, and some pesticides, Lyon7 IARC, vol. 53.

Inui, M., Adachi, T., Takenaka, S., Inui, H., Nakazawa, M., Ueda, M., Watanabe, H., Mori, C., Iguchi, T. and Miyatake, K. (2003) Effect of UV screens and preservatives on vitellogenin and choriogenin production in male medaka (Oryzias latipes). Toxicol. **194**, 43–50.

Isidori, M., Bellotta, M., Cangiano, M. and Parella, A. (2009) Estrogenic activity of pharmaceuticals in the aquatic environment. *Environ. Int.* **35**, 826–829.

Isidori, M., Nardelli, A., Pascarella, L., Rubino, M. and Parella, A. (2007) Toxic and genotoxic impact of fibrates and their photoproducts on non-target organisms. *Environ. Int.* **33**, 635–641.

John, D. M., House, W. A. and White, G. F. (2000) Environmental fate of nonylphenol ethoxylates: differential adsorption of homologs to components of river sediment. *Environ. Toxicol. Chem.* **19**, 293–300.

Jonkers, N., Kohler, H. P., Dammshauser, A. and Giger, W. (2009) Mass flows of endocrine disruptors in the Glatt River during varying weather conditions. *Environ. Pollut.* **157**, 714–723.

Joss, A., Keller, E., Alder, A. C., Gobel, A., McArdell, C. S., Ternes, T. and Siegrist, H. (2005) Removal of pharmaceuticals and fragrances in biological wastewater treatment. *Water Res.* **39**, 3139–3152.

JRC (Joint Research Center) (2001), Organic Contaminants in sewage sludge for agricultural use (http://ec.europa.eu/environment/ waste/sludge/pdf/ organics_in_ ludge.pdf, retrieved 05.11.2009).

Jürgens, M. D., Williams, R. J. and Johnson, A. C. (1999) R&D Technical Report P161, Environment Agency, Bristol, UK.

Jürgens, M. D., Holthaus, K. I. E., Johnson, A. C., Smith, J. J. L., Hetheridge, M. and Williams, R. J. (2002) The potential for estradiol and ethinylestradiol degradation in English rivers. *Environ. Toxicol. Chem.* **21**, 480–488.

Kalantzi, O. L., Martin, F. L., Thomas, G. O., Alcock, R. E., Tang, H. R., Drury, S. C., Carmichael, P. L., Nicholson, J. K. and Jones, K. C. (2004) Different levels of polybrominated diphenyl ethers (PBDEs) and chlorinated compounds in breast milk from two U.K. regions. *Environ. Health Persp.* **112**, 1085–1091.

Kannan, K., Corsolini, S., Falandysz, J., Fillmann, G., Kumar, K. S., Loganathan, B. G., Mohd, M. A., Olivero, J., Van Wouwe, N., Yang, J. H. and Aldoust, K. M. (2004) Perfluorooctanesulfonate and related fluorochemicals in human blood from several countries. *Environ. Sci. Technol.* **38**, 4489–4495.

Kannan, K., Tao, L., Sinclair, E., Pastva, S. D., Jude, D. J. and Giesy, J. R. (2005) Perfluorinated compounds in aquatic organisms at various trophic levels in a Great Lakes food chain. *Arch. Environ. Cont. Toxicol.* **48**, 559–566.

Kasprzyk-Hordern, B., Dinsdale, R. M. and Guwy, A. J. (2007) Multi-residue method for the determination of basic/neutral pharmaceuticals and illicit drugs in surface water by solid-phase extraction and ultra performance liquid chromatography–positive electrospray ionisation tandem mass spectrometry. *J. Chromatogr. A* **1161**, 132–145.

Kasprzyk-Hordern, B., Dinsdale, R. M. and Guwy, A. J. (2008) The occurrence of pharmaceuticals, personal care products, endocrine disruptors and illicit drugs in surface water in South Wales, UK. *Water Res.* **42**, 3498–3518.

Kaushik, P. and Kaushik, G (2007) An assessment of structure and toxicity correlation in organochlorine pesticides. *J. Hazard. Mater.* **143**, 102–111.

Kelce, W. R. and Wilson, E. M. (1997) Environmental antiandrogens: developmental effects, molecular mechanisms, and clinical implications. *J. Mol. Med.* **75**, 198–207.

Keller, J. M., Kannan, K., Taniyasu, S., Yamashita, N., Day, R. D., Arendt, M. D., Segars, A. L. and Kucklick, J. R. (2005) Perfluorinated compounds in the plasma of loggerhead and Kemp's ridley sea turtles from the southeastern coast of the United States. *Environ. Sci. Technol.* **39**, 9101–9108.

Kemper, R. A. (2003) Perfluorooctanoic acid: Toxicokinetics in the rat. DuPont Haskell Laboratories, Project No. DuPont–7473. US EPA Administrative Record, AR-226-1499.

Kim, S.-C. and Carlson, K. (2007) Temporal and spatial trends in the occurrence of human and veterinary antibiotics in aqueous and river sediment matrices. *Environ. Sci. Technol.* **41**, 50–57.

Kissa, E. (2001) Fluorinated Surfactants and Repellents (second ed), Marcel Dekker, Inc., New York, NY, USA.

Kolpin, D. W., Furlong, E. T., Meyer, M. T., Thurman, E. M., Zaugg, S. D., Barber, L. B. and Buxton, H. T. (2002) Pharmaceuticals, hormones and other organic wastewater contaminants in U.S. Streams, 1999–2000: A National Reconnaissance. *Environ. Sci. Technol.* **36**, 1202–1211.

Koppen, G., Covaci, A., Van Cleuvenbergen, R., Schepens, P., Winneke, G., Nelen, V., van Larebeke, N., Vlietinck, R. and Schoeters, G. (2008) Persistent organochlorine pollutants in human serum of 50–65 years old women in the Flanders Environmental and Health Study (FLEHS). Part 1: concentrations and regional differences. *Chemosphere* **48**, 811–825.

Kreuzinger, N., Clara, M., Strenn, B. and Kroiss, H. (2004) Relevance of the sludge retention time (SRT) as design criteria for wastewater treatment plants for the removal of endocrine disruptors and pharmaceuticals from wastewater. *Water Sci. Technol.* **50**, 149–156.

Kuch, H.M. and Ballschmiter, K. (2001) Determination of endocrine disrupting phenolic compounds and estrogens in surface and drinking water by HRGC-(NCI)-MS in the picogram per liter range. *Environ. Sci. Technol.* **35**, 3201–3206.

Kumar, V., Chakraborty, A., Kural, M. R. and Poy, P. (2009) Alteration of testicular steroidogenesis and histopathology of reproductive system in male rats treated with triclosan. *Reprod. Toxicol.* **27**, 177–185.

Kummerer, K. (2009) The presence of pharmaceuticals in the environment due to human use – present knowledge and future challenges. *J. Environ. Manage.* **90**, 2354–2366.

Kunz, P. Y. and Fent, K. (2006) Estrogenic activity of UV filter mixtures. *Toxicol. Appl. Pharm.* **217**, 86–99.

Lagana, A., Bacaloni, A., De Leva, I., Faberi, A., Fago, G. and Marino, A. (2004) Analytical methodologies for determining the occurrence of endocrine disrupting chemicals in sewage treatment plants and natural waters. *Anal. Chim. Acta* **501**, 79–88.

Lai, K. M., Scrimshaw, M. D. and Lester, J. N. (2002) Biotransformation and bioconcentration of steroid estrogens by *Chlorella vulgaris*. *App. Environ. Microb.* **68**, 859–864.

Lai, K. M., Johnson, K. L., Scrimshaw, M. D. and Lester, J. N. (2000) Binding of waterborne steroid estrogens to solid phases in river and estuarine systems. *Environ. Sci. Technol.* **34**, 3890–3894.

Lalah, J. D., Schramm, K. W., Henkelmann, B., Lenoir, D., Behechti, A., Gunther, K. and Kettrup, A. (2003) The dissipation, distribution and fate of a branched ^{14}C-nonylphenol isomer in lake water/sediment systems. *Environ. Pollut.* **122**, 195–203.

Lam, M. and Mabury, S. A. (2005) Photodegradation of the pharmaceuticals atorvastatin, carbamazepine, levofloxacin, and sulfamethoxazole in natural waters. *Aquat. Sci.* **67**, 177–188.

Lam, M. W., Young, C. J., Brain, R. A., Johnson, D. J., Hanson, M. A., Wilson, C. J., Richards, S. M., Solomon, K. R. and Mabury, S. A. (2004) Aquatic persistence of eight pharmaceuticals in a microcosm study. *Environ. Toxicol. Chem.* **23**, 1431–1440.

Lambropoulou, D. A., Sakkas, V. A., Hela, D. G. and Albanis, T. A. (2002) Application of solid phase microextraction (SPME) in monitoring of priority pesticides in Kalamas River (N.W. Greece). *J. Chromatogr. A* **963**, 107–116.

Lange, I. G., Daxenberger, A., Schiffer, B., Witters, H., Ibarreta, D. and Meyer, H. H. (2002) Sex hormones originating from different livestock production systems: fate and potential disrupting activity in the environment. *Anal. Chim. Acta* **473**, 27–37.

Länge, R., Hutchinson, T. H., Croudace, C. P. and Siegmund, F. (2001) Effects of the synthetic estrogen 17 alpha-ethinylestradiol on the life-cycle of the fathead minnow (Pimephales promelas). *Environ. Toxicol. Chem.* **20**, 1216–1227.

Larsson, D. G. J., Adolfsson-Erici, M., Parkkonen, J., Pettersson, M., Berg, A. H., Olsson, P. E. and Förlin, L. (1999) Ethinyloestradiol: an undesired fish contraceptive? *Aquat. Toxicol.* **45**, 91–97.

Latch, D. E., Packer, J. L., Stender, B. L., VanOverbeke, J., Arnold, W. A. and McNeill, K. (2005) Aqueous photochemistry of triclosan: formation of 2,4-dichlorophenol, 2,8-dichlorodibenzo-p-dioxin and oligomerization products. *Environ. Toxicol. Chem.* **24**, 517–525.

Lau, C., Anitole, K., Hodes, C., Lai, D., Pfahles-Hutchens, A. and Seed, J. (2007) Perfluoroalkyl acids: A review of monitoring and toxicological findings. *Toxicol. Sci.* **99**, 366–394.

Lau, C., Butenhoff, J. L. and Rogers, J. M. (2004) The developmental toxicity of perfluoroalkyl acids and their derivatives. *Toxicol. Appl. Pharmacol.* **198**, 231–241.

Leclercq, M., Mathieu, O.,Gomez, E., Casellas, C., Fenet, H. and Hillaire-Buys, D. (2009) Presence and fate of carbamazepine, oxcarbazepine, and seven of their metabolites at wastewater treatment plants. *Arch. Environ. Cont. Toxicol.* **56**, 408–415.

Lehmler, H. J. (2005) Synthesis of environmentally relevant fluorinated surfactants – a review. *Chemosphere* **58**, 1471–1496.

Lerner, D. T., Bjornsson, B. T. and Mccormick, S. D. (2007) Larval exposure to 4-nonylphenol and 17β-estradiol affects physiological and behavioral development of seawater adaptation in Atlantic salmon smolts. *Environ. Sci. Technol.* **41**, 4479–4485.

Li, M. H. (2009) Toxicity of perfluorooctane sulfonate and perfluorooctanoic acid to plants and aquatic invertebrates. *Environ. Toxicol.* **24**, 95–101.

Lin, A. Y. and Reinhard, M. (2005) Photodegradation of common environmental pharmaceuticals and estrogens in river water. *Environ. Toxicol. Chem.* **24**, 1303–1309.

Liu, J., Wang, L., Zheng, L., Wang, X. and Lee, F. S. C. (2006) Analysis of bacteria degradation products of methyl parathion by liquid chromatography/electrospray time-of-flight mass spectrometry and gas chromatography/mass spectrometry. *J. Chromatogr. A* **29**, 180–187.

Liu, Z.-H., Kanjo, Y. and Mizutani, S. (2009) Urinary excretion rates of natural estrogens and androgens from humans, and their occurrence and fate in the environment: A review. *Sci. Total Environ.* **407**, 4975–4985.

Löffler, D., Römbke, J., Meller, M. and Ternes, T. A. (2005) Environmental fate of pharmaceuticals in water/sediment systems. *Environ. Sci. Technol.* **39**, 5209–5218.

Loos, R., Gawlik, B. M., Locoro, G., Rimaviciute, E., Contini, S. and Bidoglio, G. (2009) EU-wide survey of polar organic persistent pollutants in European river waters. *Environ. Pollut.* **157**, 561–568.

Loos, R., Wollgast, J., Huber, T. and Hanke, G. (2007) Polar herbicides, pharmaceutical products, perfluorooctanesulfonate (PFOS), perfluorooctanoate (PFOA), and nonylphenol and its carboxylates and ethoxylates in surface and tap waters arround Lake Maggiore inn Northern Italy. *Anal. Bioanal. Chem.* **387**, 1469–1478.

Lopez-Avila, V. and Hites, R. A. (1980) Organic compounds in an industrial wastewater. Their transport into sediment. *Environ. Sci. Technol.* **14**, 1382–1390.

Luo, X., Mai, B., Yang, Q., Fu, J., Sheng, G. and Wang, Z., (2004) Polycyclic aromatic hydrocarbons (PAHs) and organochlorine pesticides in water columns from the Pearl River and the Macao harbor in the Pearl River Delta in South China. *Mar. Pollut. Bull.* **48**, 1102–1115.

Manahan, S. E. (2004) Environmental Chemistry, CRC Press, New York, USA.

Martin, J. W., Mabury, S. A., Solomon, K. R. and Muir, D. C. G. (2003) Bioconcentration and tissue distribution of perfluorinated acids in rainbow trout (*Oncorhynchus Mykiss*). *Environ. Toxicol. Chem.* 196–204.

Matthews, G. (2006) Pesticides: health, safety and the environment, Blackwell Publishing, Oxford, UK.

Mc Avoy, D. C., Schatowitz, B., Jacob, M., Hauk, A. and Eckhoff, W. S. (2002) Measurement of triclosan in wastewater treatment systems. *Environ. Toxicol. Chem.* **21**, 1323–1329.

McMahon, K., Bengtson–Nash, S., Eaglesham, G, Müller, J. F., Duke, N. C. and Winderlich, S. (2005) Herbicide contamination and the potential impact to seagrass meadows in Hervey Bay, Queensland, Australia. *Mar. Pollut. Bull.* **51**, 325–334.

Metcalfe, C. D., Koenig, B. G., Bennie, D. T., Servos, M., Ternes, T. A. and Hirsch, R. (2003) Occurrence of neutral and acidic drugs in the effluents of Canadian sewage treatment plants. *Environ. Toxicol. Chem.* **22**, 2872–2880.

Mezcua, M., Gomez, M. J., Ferrer, I., Aguera, A., Hernando, M. D. and Fernandez-Alba, A. R. (2004) Evidence of 2,7/2,8-dibenzodichloro-p-dioxin as a photodegradation product of triclosan in water and wastewater samples. *Anal. Chim. Acta* **524**, 241–247.

Miao, X. S. and Metcalf, C. D. (2003) Determination of carbamazepine and its metabolites in aqueous samples using Liquid Chromatography−Electrospray Tandem Mass Spectrometry. *Anal. Chem.* **75**, 3731–3738.

Mimeault, C., Woodhouse, A. J., Miao, X. S., Metcalfe, C. D., Moon, T. W. and Trudeau, V. L. (2005) The human lipid regulator, gemfibrozil bioconcentrates and reduces testoterone in the goldfish, Carassius auratus. *Aquat. Toxicol.* **73**, 44–54.

Mompelat, S., Le Bot, B. Thomas, O. (2009) Occurrence and fate of pharmaceutical products and by-products, from resource to drinking water. *Environ. Int.* **35**, 803–814.

Müller, B., Berg, M., Yao, Z.-P., Zhang, X.-F., Wang, D. and Pfluger, A. (2008) How polluted is the Yangtze river? Water quality downstream from the Three Gorges Dam. *Sci. Total Environ.* **402**, 232–247.

Murakami, M., Shinohara, H. and Takada, H. (2009) Evaluation of wastewater and street runoff as sources of perfluorinated surfactants (PFSs). *Chemosphere* **74**, 487–493.

Nelson, S. D., Letey, J., Farmer, W. J. and Ben-Hur, M. (1998) Facilitated transport of nanpropamide by dissolved organic matter in sewage sludge-amended soil. *J. Environ. Qual.* **27**, 1194–2000.

Ning, B., Graham, N., Zhang, Y., Nakonechny, M. and El-Din, M. G. (2007) Degradation of Endocrine Disrupting Chemicals by Ozone/AOPs. *Ozone-Sci. Eng.* **29**, 153–176.

Oaks, J. L., Gilbert, M., Virani, M. Z., Watson, R. T., Meteyer, C. U., Rideout, B. A., Shivaprasad, H. L., Ahmed, S., Chaudhry, M. J. I., Arshad, M., Mahmood, S., Ali, A. and Khan, A. A. (2004) Diclofenac residues as the cause of vulture population decline in Pakistan. *Nature* **427**, 630–633.

OECD (Organization for Economic Co-operation and Development), 2005. Results of survey on production and use of PFOS, PFAS and PFOA, related substances and products/mixtures containing these substances. ENV/JM/MONO(2005)1, Paris.

Okamura, H., Aoyama, I., Ono, Y. and Nishida, T. (2003) Antifouling herbicides in the coastal waters of western Japan, *Mar. Pollut. Bull.* **47**, 59–67.

Olsen, G. W., Burris, J. M., Ehresman, D. J., Froehlich, J. W., Seacat, A. M., Butenhoff, J. L. and Zobel, L. R. (2007) Half-life of serum elimination of perfluorooctanesulfonate, perfluorohexanesulfonate, and perfluorooctanoate in retired fluorochemical production workers. *Environ. Health Perspect.* **115**, 1298–1305.

Orvos, D. R., Versteeg, D. J., Inauen, J., Capdevielle, M., Rodethenstein, A. and Cunningham, V. (2002) Aquatic toxicity of triclosan. *Environ. Toxicol. Chem.* **21**, 1338–1349.

Palmer, B. D. and Palmer, S. K. (1995) Vitellogenin induction by xenobiotic estrogens in the red-eared turtle and African clawed frog. Environ. *Health Perspect.* **103**, 19–25.

Pandit, G. G., Mohan Rao, A. M., Jha, S. K., Krishnamoorthy, T. M., Kale, S. P., Raghu, K. and Murthy, N. B. K. (2001) Monitoring of organochlorine pesticide residues in the Indian marine environment. *Chemosphere* **44**, 301–305.

Pehkonen, S.O. and Zhang, Q. (2002) The degradation of organophosphorus pesticides in natural waters: a critical review. *Crit. Rev. Env. Sci. Tec.* **32**, 17–72.

Peng, X., Yu, Y., Tang, C., Tan, J. Xuang, Q. and Wang, Z. (2008) Occurrence of steroid estrogens, endocrine-disrupting phenols, and acid pharmaceutical residues in urban riverine water of the Pearl River Delta, South China. *Sci. Total Environ.* **397**, 158–166.

Peñuela, G. A. and Barceló, D. (1998) Photosensitized degradation of organic pollutants in water: processes and analytical applications. *Trend. Anal. Chem.* **17**, 605–612.

Petrovic, M., Diaz, A., Ventura, F. and Barcelo, D. (2003) Occurrence and removal of estrogenic short-chain ethoxy nonylphenolic compounds and their halogenated derivatives during drinking water production. *Environ. Sci. Technol.* **37**, 4442–4448.

Planas, C., Caixach, J., Santos, F. J. and Rivera, J. (1997) Occurrence of pesticides in Spanish surface waters. Analysis by high resolution gas chromatography coupled to mass spectrometry. *Chemosphere* **34**, 2393–2406.

Planas, C., Guadayol, J. M., Droguet, M., Escalas, A., Rivera, J. and Caixach, J. (2002) Degradation of polyethoxylated nonylphenols in a sewage treatment plant. Quantitative analysis by isotopic dilution-HRGC/MS. *Water Res.* **36**, 982–988.

Plumlee, M. H., Larabee, J. and Reinhard, M. (2008) Perfluorochemicals in water reuse. *Chemosphere* **72**, 1541–1547.

Poiger, T., Buser, H.-R. and Müller, M. D. (2001) Photodegradation of the pharmaceutical drug diclofenac in a lake: pathway, field measurements, and mathematical modeling. *Environ. Toxicol. Chem.* **20**, 256–263.

Polder, A., Odland, J. O., Tkachev, A., Foreid, S., Savinova, T. N. and Skaare, J. U. (2003) Geographical variation of chlorinated pesticides, toxaphenes and PCBs in human milk from sub-arctic and arctic locations in Russia. *Sci. Total Environ.* **306**, 79–195.

Pomati, F., Orlandi, C., Clerici, M., Luciani, F. and Zuccato, E. (2008) Effects and interactions in an environmentally relevant mixture of pharmaceuticals. *Toxicol. Sci.* **102**, 129–137.

Preuss, T. G., Gurer-Orham, H., Meerman, J. and Ratte, H. T. (2009) Some nonylphenol isomers show antiestrogenic potency in the MVLN cell assay. Toxicology in Vitro (in press, doi: 10.1016/j.tiv.2009.08.017).

Purdum, C. E., Hardiman, P. A., Bye, V. J., Eno, N. C., Tyler ,C. R. and Sumpter, J. P. (1994) Oestrogenic effects of effluent from sewage treatment works. *Chem. Ecol.* **8**, 275–285.

Qiu, Y.-W., Zhang, G., Guo, L.-L., Cheng, H.-R., Wang, W.-X., Li, X.-D. and Wai, W. H. (2009) Current status and historical trends of organochlorine pesticides in the ecosystem of Deep Bay, South China. *Estuar. Coast. Shelf S.* **85**, 265–272.

Quinn, B., Gagné, F. and Blaise, C. (2009) Evaluation of the acute, chronic and teratogenic effects of a mixture of eleven pharmaceuticals on the cnidarian, Hydra attenuate. *Sci. Total Environ.* **407**, 1072–1079.

Radjenovic, J., Petrovic, M. and Barcelo, D. (2009) Fate and distribution of pharmaceuticals in wastewater and sewage sludge of the conventional activated sludge (CAS) and advanced membrane bioreactor (MBR) treatment. *Water Res.* **43**, 831–841.

Rana, S. V. S. (2006) Environmental Pollution, Health and Toxicology, Alpha Science International Lts., Oxford, UK.

Reith, D. M., Appleton, D. B., Hooper, W. and Eadie, M. J. (2000) The effect of body size on the metabolic clearance of carbamazepine. *Biopharm. Drug Dispos.* **21**, 103–111.

Renner, R. (1997) European bans on surfactant trigger transatlantic debate. *Environ. Sci. Technol.* **31**, 316–320.

Rodgers-Gray, T. P., Jobling, S., Morris, S., Kelly, C., Kirby, S., Janbakhsh, A., Harries, J. E., Waldock, M. J., Sumpter, J. P. and Tyler, C. R. (2000) Long-term temporal changes in the estrogenic composition of treated sewage effluent and its biological effects on fish. *Environ Sci Technol.* **34**, 1521–1528.

Rupa, D. S., Reddy, P. P. and Reddi, O. S. (1990) Cytogeneticity of quinalphos and methyl parathion in human peripheral lymphocytes. *Hum. Exp. Toxicol.* **9**, 385–387.

Sabik, H., Jeannot, R. and Rondeaua, B. (2000) Multiresidue methods using solid-phase extraction techniques for monitoring priority pesticides, including triazines and degradation products, in ground and surface waters. *J. Chromatogr. A*, **885**, 217–236.

Salazar-Arredondo, E., Solís-Heredia, M. de, J., Rojas-García, E., Hernández-Ochoa, I. and Betzabet Quintanilla-Vega, B. (2008) Sperm chromatin alteration and DNA damage by methyl-parathion, chlorpyrifos and diazinon and their oxon metabolites in human spermatozoa. *Reprod. Toxicol.* **25**, 455–460.

Salvestrini, S., Di Cerbo, P. and Capasso, S. (2002) Kinetics of the chemical degradation of diuron. *Chemosphere* **48**, 69–73.

Sanderson, H., Boudreau, T. M., Mabury, S. A. and Solomon, K. R. (2004) Effects of perfluorooctane sulfonate and perfluorooctanoic acid on the zooplanktonic community. *Ecotox. Environ. Safe.* **58**, 68–76.

Santos, L., Aparicio, I. and Alonso, E. (2007) Occurrence and risk assessment of pharmaceutically active compounds in wastewater treatment plants. A case study: Seville city (Spain). *Environ. Int.* **33**, 596–601.

Sarmah, A. K., Meyer, M. T. and Boxall, A. B. A. (2006) A global perspective on the use, sales, exposure pathways, occurrence, fate and effects of veterinary antibiotics (Vas) in the environment. *Chemosphere* **65**, 725–759.

Sarmah, A. K. and Northcott, G. L. (2008) Laboratory degradation studies of four endocrine disruptors in two environmental media. *Environ. Toxicol. Chem.* **27**, 819–827.

Scheytt, T., Mersmann, P., Lindstädt, R. and Heberer, T. (2005) Determination of pharmaceutically active substances carbamazepine, diclofenac, and ibuprofen, in sandy sediments. *Chemosphere* **60**, 245–253.

Schneider, W. and Degen, P. H. (1981) Simultaneous determination of diclofenac sodium and its hydroxy metabolites by capillary column gas chromatography with electron-capture detection. *J. Chromatogr.* **217**, 263–271.

Schubert, S., Peter, A., Burki, R., Schonenberger, R., Suter, M. J. F. and Segner, H., Burkhardt-Holm P. (2008) Sensitivity of brown trout reproduction to long-term estrogenic exposure. *Aquat. Toxicol.* **90**, 65–72.

Schwaiger, J., Ferling, H., Mallow, U., Wintermayr, H. and Negele, R. D. (2004) Toxic effects of the non-steroidal anti-inflammatory drug diclofenac. Part I: Histopathological alterations and bioaccumulation in rainbow trout. *Aquat. Toxicol.* **68**, 141–150.

Seacat, A. M., Thomford, P. J., Hansen, K. J., Olsen, G. W., Case, M. T. and Butenhoff, J. L. (2002) Subchronic toxicity studies on perfluorooctanesulfonate potassium salt in cynomolgus monkeys. *Toxicol. Sci.* **68**, 249–264.

Segura, P. A., Francois, M., Cagnon, C. and Sauve, S. (2009) Review of the occurrence of anti-infectives in contaminated wastewaters and natural and drinking waters. *Environ. Health Persp.* **117**, 675–684.

Seibert, B. (1996) Data from animal experiments and epidemiology data on tumorigenicity of estradiol valerate and ethinyl estradiol, cited in: Endocrinically Active Chemcials in the Environment, UBA TEXTE 3/96, Berlin, pp. 88–95.

Servos, M. R., Bennie, D. T., Burnison, B. K., Jurkovic, A., McInnis, R., Neheli, T., Schnell, A, Seto, P., Smyth, S. A. and Ternes, T. A. (2005) Distribution of estrogens, 17b-estradiol and estrone, in Canadian municipal wastewater treatment plants. *Sci. Total Environ.* **336**, 155–170.

Shankar, M. V., Nélieu, S., Kerhoas, L. and Einhorn, J. (2008) Natural sunlight $NO^{3\ -}$/ NO^{2-}-induced photo-degradation of phenylurea herbicides in water. *Chemosphere* **71**, 1461–1468.

Shao, B., Hu, J., Yang, M., An, W. and Tao, S. (2005) Nonylphenol and nonylphenol ethoxylates in river water, drinking water, and fish Tissues in the area of Chongqing, China. *Arch. Environ. Cont. Toxicol.* **48**, 467–473.

Shore, L. S., Kapulnik, Y., Ben-Dov, B., Fridman, Y., Wininger, S. and Shemesh, M. (1992) Effects of estrone and 17 β estradiol on vegetative growth of Medicago sativa, Physiol. *Plant.* **84**, 217–222.

Shukla, G., Kumar, A., Bhanti, M., Joseph, P. E. and Taneja, A. (2006) Organochlorine pesticide contamination of ground water in the city of Hyderabad. *Environ. Int.* **32**, 244–247.

Shultz, S., Baral, H. S., Charman, S., Cunningham, A. A., Das, D., Ghalsasi, G. R., Goudar, M., Green, R. E., Jones, A., Nighot, P., Pain, D. J. and Prakash, V. (2004) Diclofenac poisoning is widespread in declining vulture populations across the Indian subcontinent. *Proc. Roy. Soc. London B* **271**, S458–S460.

Sinclair, E. and Kannan, K. (2006) Mass loading and fate of perfluoroalkyl surfactants in wastewater treatment plants. *Environ. Sci. Technol.* **40**, 1408–1414.

Singer, H., Muller, S., Tixier, C. and Pillonel, L. (2002) Triclosan: Occurrence and fate of a widely used biocide in the aquatic environment: Field measurements in wastewater treatment plants, surface waters, and lake sediments. *Environ. Sci. Technol.* **36**, 4998–5004.

Soares, A., Guieysse, B., Jefferson, B., Cartmell, E. and Lester, J. N. (2008) Nonylphenol in the environment: A critical review on occurrence, fate, toxicity and treatment in wastewaters. *Environ. Int.* **34**, 1033–1049.

Soto, A. M., Sonnenschein, C., Chung, K. L., Fernandez, M. F., Olea, N. and Serrano, F. O. (1995) The E-SCREEN assay as a tool to identify estrogens: an update on estrogenic environmental pollutants. *Environ. Health Persp.* **103**, 113–122.

Standley, L. J., Rudel, R. A., Swartz, C. H., Attfield, K. R., Christian, J., Erickson, M. and Brody, J. G. (2008) Wastewater-contaminated groundwater as a source of endogenous hormones and pharmaceuticals to surface water ecosystems. *Environ. Tox. Chem.* **27**, 2457–2468.

Staples, C., Mihaich E., Carbone J., Woodburn K., Klecka G. (2004) A weight of evidence analysis of the chronic ecotoxicity of nonylphenol ethoxylates, nonylphenol ether carboxylates, and nonylphenol. *Hum. Ecol. Risk Assess.* **10**, 999–1017.

Stasinakis, A. S., Gatidou, G., Mamais, D., Thomaidis, N. S. and Lekkas, T. D. (2008) Occurrence and fate of endocrine disrupters in Greek sewage treatment plants. *Water Res.* **42**, 1796–1804.

Stasinakis, A. S., Kordoutis, C., I., Tsiouma, V. C., Gatidou, G. and Thomaidis, N. S. (2009b) Removal of selected endocrine disrupters in activated sludge systems: Effect of sludge retention time on their sorption and biodegradation. Bioresource Technol. (in press).

Stasinakis, A. S., Petalas, A. V., Mamais, D., Thomaidis, N. S., Gatidou, G. and Lekkas, T. D. (2007) Investigation of triclosan fate and toxicity in continuous-flow activated sludge systems. *Chemosphere*, **68**, 375–381.

Stasinaksi, A. S., Kotsifa, S., Gatidou, G., Mamais, D. (2009a) Diuron biodegradation in activated sludge batch reactors under aerobic and anoxic conditions. *Water Res.* **43**, 1471–1479.

Strandberg, M. T. and Scott-Fordsmand, J.-J. (2002) Field effects of simazine at lower trophic levels–a review. *Sci. Total Environ.* **296**, 117–137.

Takagi, S., Adachi, F., Miyano, K., Koizumi, Y., Tanaka, H., Mimura, M., Watanabe, I., Tanabe, S. and Kannan, K. (2008) Perfluorooctanesulfonate and perfluorooctanoate in raw and treated tap water from Osaka, Japan. *Chemosphere* **72**, 1409–1412.

Tao, L., Kannan, K., Kajiwara, N., Costa, M., Fillman, G., Takahashi, S. and Tanabe, S. (2006) Perfluorooctanesulfonate and related flurochemicals in albatrosses, elephant seals, penguins, and polar skuas from the Southern Ocean. *Environ. Sci. Technol.* **40**, 7642–7648.

Ternes, T. A., Hirsch, R., Mueller, J. and Haberer, K. (1998) Methods for the determination of neutral drugs as well as betablockers and β_2-sympathomimetics in aqueous matrices using GC/MS and LC/MS/MS. Fresen. *J. Anal. Chem.* **362**, 329–340.

Ternes, T. A., Knacker, T. and Oehlmann, J. (2003) Persconal care products in the aquatic environment – A group of substances which has been neglected to date. Umweltwissenschaften und Schadstoff-Forschung, **15**, 169–180.

Thiele-Bruhn, S. (2003) Pharmaceutical antibiotic compounds in soils – a review. *J. Plant. Nutr. Soil Sci.* **166**, 145–167.

Thomas, D. B. (1984), Do hormones cause breast cancer? *Cancer* **53**, 595–604.

Thomas, L., Russell, A. D. and Maillard, J. Y. (2005) Antimicrobial activity of chlorhexidine diacetate and benzalkonium chloride against Pseudomonas aeruginosa and its response to biocide residues. *J. Appl. Microb.* **98**, 533–543.

Thorpe, K. L., Hutchinson, T. H., Hetheridge, M. J., Scholze, M., Sumpter, J. P. and Tyler, C. R. (2001) Assessing the biological potency of binary mixtures of environmental estrogens using vitellogenin induction in juvenile rainbow trout (Oncorhynchus mykiss). *Environ. Sci. Technol.* **35**, 2476–2481.

Thorpe, K. L., Cummings, R. I., Hutchinson, T. H., Scholze, M., Brighty, G., Sumpter, J. P. Tyler, C. R. (2003) Relative potencies and combination effects of steroidal estrogens in fish. *Environ Sci Technol* **37**, 1142–1149.

Tixier, C., Singer, H. P., Canonica, S. and Muller, S. R. (2002) Phototransformation of triclosan in surface waters: a relevant elimination process for this widely used biocide – Laboratory studies, field measurements, and modeling. *Environ. Sci. Technol.* **36**, 3482–3489.

Tolls, J. (2001) Sorption of veterinary pharmaceuticals in soils: a review. *Environ. Sci. Technol.* **17**, 3397–3406.

Tomy, G. T., Tittlemier, S. A., Palace, V. P., Budakowski, W. R., Brarkevelt, E., Brinkworth, L. and Friesen, K. (2004) Biotransformation of N-ethyl perfluorooctanesulfonamide by rainbow trout (Onchorhynchus mykiss) liver microsomes. *Environ. Sci. Technol.* **38**, 758–762.

Triebskorn, R., Casper, H., Heyd, A., Eikemper, R., Köhler, H.-R. and Schwaiger, J. (2004) Toxic effects of the non-steroidal anti-inflammatory drug diclofenac. Part II: Cytological effects in liver, kidney, gills and intestine of rainbow trout (Oncorhynchus mykiss). *Aquat. Toxicol.* **68**, 151–166.

U.S. Environmental Protection Agency: http://www.epa.gov/opp00001/about/ (retrieved 20.10.2009).

U.S. EPA (U.S. Environmental Protection Agency) (2000) Water quality standards: establishment of numeric criteria for priority toxic pollutants for the State of California; final rule. *Fed. Reg.* **65**, 31681–31719.

UNEP (2002) Sub-Saharan Africa Regional Report: Regionally Based Assessment of Persistent Toxic Substances.

UNEP (2003) Stockholm Convention: Master List of Actions: on the reduction and/or elimination of the releases of persistent organic pollutants (Fifth ed.), United Nations Environmental Programme, Geneva, Switzerland.

Van Aerle, R., Rounds, N., Hutchinson, T. H., Maddix, S. and Tyler, C. R. (2002) Window of sensitivity for the estrogenic effects of ethinylestradiol in early life-stages of fathead minnow. *Ecotoxicology* **11**, 423–434.

Vanderford, B. J., Pearson, R. A., Rexing, D. J. and Snyder, S.A. (2003) Analysis of endocrine disruptors, pharmaceuticals and personal care products in water using Liquid Chromatography/Tandem Mass Spectrometry. *Anal. Chem.* **75**, 6265–6274.

Vazquez-Duhalt, R., Marquez-Rocha, F., Ponce, E., Licea, A. F. and Viana, M. T. (2005) Nonylphenol, an integrated vision of a pollutant. Scientific review. *Appl. Ecol. Environ. Res.* **4**, 1–25.

Veldhoen, N., Skirrow, R. C., Osachoff, H., Wigmore, H., Clapson, D. J., Gunderson, M. P., Van Aggelen, G. and Helbing, C. C. (2006) The bactericidal agent triclosan modulates thyroid hormone-associated gene expression and disrupts postembryonic anuran development. *Aquat. Toxicol.* **80**, 217–227.

Vestergren, R. and Cousins, I. T. (2009) Tracking the pathways of human exposure to perfluorocarboxylates. *Environ. Sci. Technol.* **43**, 5565–5575.

Vorkamp, K., Riget, F., Glasius, M., Pécseli, M., Lebeufand, M. and Muir, D. (2004) Chlorobenzenes, chlorinated pesticides, coplanar chlorobiphenyls and other organochlorine compounds in Greenland biota. *Sci. Total Environ.* **331**, 157–175.

Walker, C. H., Hopkin, S. P., Silby, R. M. and Peakall, D. B. (2006) Principles of

Ecotoxicology, CRC Press, New York, USA.

Wang, N., Stostek, B., Folsom, P. W., Sulecki, L. M., Capka, V., Buck, R. C., Berti, W. R. and Gannon, J. T. (2005) Aerobic biotransformation of 14C-labeled 8-2 telomer B alcohol by activated sludge from domestic sewage treatment plant. *Environ. Sci. Technol.* **39**, 531–538.

Weiger, S., Berger, U., Jensen, E., Kallenborn, R., Thoresen, H. and Huhnerfuss, H. (2004) Determination of selected pharmaceuticals and caffeine in sewage and seawater from Tromsø/Norway with emphasis on ibuprofen and its metabolites. *Chemosphere*, **56**, 583–592.

WHO (World Health Organization) (2004) The WHO recommended classification of pesticides by hazard and guidelines to classification. WHO, Geneva.

Winker, M., Faika, D., Gulyas, H. and Otterpohl, R. (2008) A comparison of human pharmaceutical concentrations in raw municipal wastewater and yellowwater. *Sci. Total Environ.* **399**, 96–104.

Winkler, M., Lawrence, J. R. and Neu, T. R. (2001) Selective degradation of ibuprofen and clofibric acid in two model river biofilm systems. *Water Res.* **35**, 3197–3205.

Worthing, C. R. and Walker, S. B. (1987) The Pesticide Manual: a World Compendium (eight ed.), British Crop Protection Council, London.

Wu, X., Hua, R., Tang, F., Li, X., Cao, H. and Yue, Y. (2006) Photochemical degradation of chlorpyrifos in water. *Chinese J. App. Ecol.* **17**, 1301–1304.

Yamamoto, H., Nakamura, Y., Moriguchi, S., Nakamura, Y., Honda, Y., Tamura, I., Hirata, Y., Hayashi, A. and Sekizawa, J. (2009) Persistence and partitioning of eight selected pharmaceuticals in the aquatic environment: Laboratory photolysis, biodegradation, and sorption experiments. *Water Res.* **43**, 351–362.

Yamamoto, H., Watanabe, M., Katsuki, S., Nakamura, Y., Moriguchi, S., Nakamura, Y. and Sekizawa, J. (2007) Preliminary ecological risk assessment of butylparaben and benzylparaben-2. Fate and partitioning in aquatic environments. *Environ. Sci. : An Int. J. Environ. Phys. Toxicol.* **14**, 97–105.

Yamashita, N., Kannan, K., Taniyasu, S., Horii, Y., Petrick, G. and Gamo, T. (2005) A global survey of perfluorinated acids in oceans. *Mar. Pollut. Bull.* **51**, 658–668.

Ye, X., Bishop, A. M., Reidy, J. A., Needham, L. L. and Calafat, A. M. (2006) Parabens as urinary biomarkers of exposure in humans. *Environ. Health Persp.* **114**, 1843–1846.

Ying, G. G. (2006) Fate, behavior and effects of surfactants and their degradation products in the environment. *Environ. Int.* **32**, 417–431.

Ying, G. G., Yu, X. Y. and Kookana, R. S. (2007) Biological degradation of triclocarban and triclosan in a soil under aerobic and anaerobic conditions and comparison with environmental fate modeling. *Environ. Pollut.* **150**, 300–305.

Young, C. J., Furdui, V. I., Franklin, J., Koerner, R. M., Muir, D. C. G. and Mabury, S. A. (2007) Perfluorinated acids in arctic snow: New evidence for atmospheric formation. *Environ. Sci. Technol.* **41**, 3455–3461.

Yuan, S. Y., Yu, C. H. and Chang, B. V. (2004) Biodegradation of nonylphenol in river sediment. *Environ. Pollut.* **127**, 425–430.

Zhang, Y., Geiben, S. U. and Gal, C. (2008) Carbamazepine and diclofenac: Removal in wastewater treatment plants and occurrence in water bodies. *Chemosphere* **73**, 1151–1161.

Zhou, J. L., Maskaoui, K., Qiu, Y. W., Hong, H. S. and Wang, Z. D. (2001) Polychlorinated biphenyl congeners and organochlorine insecticides in the water column and sediments of Daya bay, China. *Environ. Pollut.* **113**, 373–384.

Zhou, R., Zhu, L., Chen, Y. and Kong, Q. (2008) Concentrations and characteristics of organochlorine pesticides in aquatic biota from Qiantang River in China. *Environ. Pollut.* **151**, 190–199.

Zimetbaum, P., Frishman, W. H. and Kahn, S. (1991) Effects of gemfibrozil and other fibric acid derivates on blood lipids and lipoproteins. *J. Clin. Pharm.* **31**, 25–37.

Zwiener, C., Seeger, S., Glauner, T. and Frimmel, F. H. (2002) Metabolites from the biodegradation of pharmaceutical residues of ibuprofen in biofilm reactors and batch experiments. *Anal. Bioanal. Chem.* **372**, 569–575.

2

Micropollutants and their Identification: An Analytical Methodology

Ekaterina V. Rokhina and Jurate Virkutyte

2.1 INTRODUCTION

Micropollutants (the harmful substances detected in micro g range in the environment) are the non-regulated contaminants, which may be potential candidates for the future regulation depending on their health effects and monitoring data regarding their occurrence. The great majority of micropollutants are presented in Table 2.1. These recently emerged substances are mostly pesticides, pharmaceuticals, personal care products (PPCPs), industrial chemicals, gasoline additives and etc.

Table 2.1 Classes of emerging contaminants

Group of emerging compounds (Category/subcategories)	Examples	Review of analytical method for the specific class of compounds
Antifoaming agents	Surfinol-104	(Richardson, 2009)
Antioxidants	2,6-Di-tert-butylphenol 4-tert-Butylphenol BHA BHQ BHT	(Henry and Yonker, 2006) (Bakker and Qin, 2006)
Artificial sweeteners	Sucrose	(Giger, 2009)
Complexing agents	DTPA EDTA NTA Oxadixyl TAED	(Richardson *et al.*, 2007)
Biocides	Triclosan Methyltriclosan Chlorophene	(Richardson, 2000)
Detergents		(Richardson, 2009)
Aromatic sulphonates		
Linear alkylbenzene sulfonates (LAS)	C10-C14-LAS C12-LAS	
Ethoxylates/carboxylates of octyl/nonyl phenols	4-Nonylphenol di-ethoxylate 4-Octylphenol di-ethoxylate	
Disinfection by-products	Bromate Cyanoformaldehyde Decabromodiphenyl ethane Hexabromocyclododecane (HBCD) NDMA	(Richardson, 2003) (Richardson *et al.*, 2007)
Flame retardants		
Brominated flame retardants	Tetrabromo bisphenol A (TBBPA) Decabromodiphenyl ethane	(Hyotylainen and Hartonen, 2002)
Polybrominated diphenylethers	Technical Decabromodiphenyl ether	

(*continued*)

Table 2.1 (*continued*)

Group of emerging compounds (Category/subcategories)	Examples	Review of analytical method for the specific class of compounds
	Technical Octabromodiphenyl ether Technical Pentabromodiphenyl ether	
Organophosphates	Tri-(dichlorisopropyl)-phosphate Triethylphosphate Tri-n-butylphosphate	(Reemtsma *et al.*, 2008)
Chlorinated paraffin	Long chain PCAs (lPCAs, C > 17) Technical PCA products	
Fragrances	Acetylcedrene Benzylacetate Camphor g-Methylionone	(Raynie, 2004)
Gasoline additives	Dialkyl ethers, Methyl-t-butyl ether (MTBE)	(Richardson, 2009)
Industrial chemicals	TCEP Triphenyl phosphine oxide	
Personal care products		(Hao *et al.*, 2007) (Kot-Wasik *et al.*, 2007)
Sun-screen agents	Benzophenone 4-Methylbenzylidene camphor Octocrylene Oxybenzone	
Insect repellents	N,N-diethyl-m-toluamide (DEET) Bayrepel	
Antiseptics	Triclosan, Chlorophene	
Carriers	Octamethylcyclote-trasiloxane (D4) Decamethylcyclopen-tasiloxane (D5)	

(*continued*)

Table 2.1 (*continued*)

Group of emerging compounds (Category/subcategories)	Examples	Review of analytical method for the specific class of compounds
	Octamethyltrisiloxane (MDM) Decamethyltetrasiloxane (MD2M) Dodecamethylpenta-siloxane (MD3M)	
Parabens (hydroxybenzoic acid esters)	Methyl-paraben Ethyl-paraben Propyl-paraben Isobutyl-paraben	
Pesticides		(Gascon *et al.*, 1997) (Kralj *et al.*, 2007) (Hernandez *et al.*, 2008) (Liu *et al.*, 2008) (Raman Suri *et al.*, 2009)
Polar pesticides and their degradation products	Amitrole Bentazone Chlorpyrifos 2,4 D Diazinon Prometon Secbumeton Terbutryn	
Other pesticides	Cypermethrin Deltamethrin Permethrin	
New pesticides	Sulfonyl urea	
Degradation products of pesticides	Desisopropylatrazine Desethylatrazine	
Plasticizers		(Chang *et al.*, 2009)
Phthalates	Benzylbutylphthalate (BBP) Diethylphthalate (DEP) Dimetylphthalate (DMP)	
Other	Bisphenol A Triphenyl phosphate	
Benzophenone derivatives	2,4-Dihydroxybenzo-phenone	

(*continued*)

Table 2.1 (*continued*)

Group of emerging compounds (Category/subcategories)	Examples	Review of analytical method for the specific class of compounds
Pharmaceuticals		(De Witte *et al.*, 2009a) (Fatta *et al.*, 2007) (Gros *et al.*, 2008) (Pavlovic *et al.*, 2007) (Radjenovic *et al.*, 2007; Radjenovic *et al.*, 2009) (Kosjek and Heath, 2008)
Antibacterial	Sulfonamides Ampicillin Ciprofloxacin Sulfamerazine	(Hernandez *et al.*, 2007) (Garcia-Galan *et al.*, 2008)
Analgesics, anti-inflammatory drugs	Codein, ibuprofene acetaminophen, acetylsalicilyc acid, diclofenac, fenoprofen	(Macia *et al.*, 2007)
Antidepressant	Tetracycline Citalopram Escitalopram Sertraline	
Antidiabetic	Glyburide Metformin	
β-blockers	Atenolol Betaxolol Carazolol Metoprolol Propranolol Sotalol	(Hernando *et al.*, 2007)
Blood viscosity agents	Pentoxifylline	
Bronchodilators	Albuterol Albuterol sulfate Clenbuterol	
Diuretic	Caffeine Furosemide Hydrochlorothiazide	
Lipid regulators	Bezafibrate Clofibric acid Etofibrate Fenofibrate Fenofibric acid Gemfibrozil	

(*continued*)

Table 2.1 (*continued*)

Group of emerging compounds (Category/subcategories)	Examples	Review of analytical method for the specific class of compounds
Sedatives, hypnotics	Acecarbromal Allobarbital Amobarbital Butalbital Hexobarbital	
Psychiatric drugs	Amitryptiline Doxepine Diazepam Imapramine Nordiazepam	
X-ray contrast agents	Diatrizoate Iohexol Iomeprol Iopamidol Iopromide	
Steroid and hormones	17-alpha-Estradiol 17-alpha-Ethinylestradiol 17-beta-Estradiol Cholesterol Diethylstilbestrol Estriol Estrone Mestranol	(Lopez de Alda *et al.*, 2003; Gabet *et al.*, 2007) (Miege *et al.*, 2009) (Streck, 2009)
Surfactants and surfactant metabolites	Alkylphenol ethoxylates 4-nonylphnol 4-octylphenol alkylphenol carboxylates	(Lopez de Alda *et al.*, 2003)
Anticorrosives		(Richardson, 2009)
Textile dyes Acid Reactive Direct		(Poiger *et al.*, 2000) (Pinheiro *et al.*, 2004) (Oliveira *et al.*, 2007) (Kucharska and Grabka, 2010) (Petroviciu *et al.*, 2010)

Several new candidates were introduced by Richardson (2009) and Giger (2009) such as sucralose (artificial sweetener also known as Splenda), antimony (leachate from polyethylene terephthalate (PET) plastic bottles), siloxanes, and musks (fragrance additives found in perfumes, lotions, sunscreens, deodorants, and laundry detergents). Importantly, these groups of contaminants do not need to persist in the environment to cause negative effects since their high transformation/removal rates can be compensated by their continuous introduction into the water environment with domestic and industrial wastewater effluents.

Moreover, occurrence, risk assessment and ecotoxicological data are not available for most of these emerging contaminants, so it is difficult to predict what health effects they may have on humans and aquatic organisms. After the discharge into the environment, emerging pollutants are subjected to biodegradation, chemical and photochemical degradation that highly contribute to their elimination. Depending on the compartment in which the synthetic chemicals are present in the environment (e.g., groundwater, surface water and/or sediments) or in the technosphere (e.g., wastewater treatment plants (WWTPs) and drinking-water facilities), different transformations can take place, sometimes producing products that differ in their environmental behavior and ecotoxicological profile (Farre *et al.,* 2008). Unfortunately, these degradation products can exhibit even greater toxicity than parent compounds (e.g., the major biodegradation product of nonylphenol ethoxylates – nonylphenol, is much more persistent than the parent compound and can mimic estrogenic properties).

Analytical methodology available for various groups of emerging contaminants is still lacking and, although methods exist to analyze each of those compounds individually, the key issue is to develop multi-residue methods in which different compound classes can be determined by a single short analysis. Therefore, constant development in the techniques and improvement of currently existing methods is mainly focused in the field of analysis of micropollutants and their transformation products. Better instrument design and a fuller understanding of the mechanics of analytical processes would enable steady improvements to be made in sensitivity, precision, and accuracy. The ultimate aim is the development of a non-destructive method, which not only saves time but leaves the sample unchanged for further examination or processing (Fifield and Kealey, 2000).

Analysis of organic pollutants in wastewater is a complex process, basically due to the range of physico-chemical and toxicological properties of compounds included in the same group, e.g., pharmaceuticals. Taking into account public concerns on environmental issues, analytical studies and the consequent use of toxic reagents and solvents have increased to a point at which they became unsustainable to continue without an environmentally friendly perspective. The strategy to develop clean and environmentally benign material and techniques can be termed as "Sustainable Analytical Procedures" (SAPs) in the frame of Green Analytical Chemistry (GAC). GAC was initiated as a search for practical alternatives to the off-line treatment of wastes and residues in order to replace polluting methodologies with the clean ones (Armenta *et al.,* 2008). In general, GAC methods are used to replace toxic reagents, to miniaturize and to automate methods, dramatically reduce the amounts of reagents consumed and wastes generated, thus reducing or even avoiding potential side effects of currently available and future analytical methods.

The focus of the Chapter is on the general trends and the most used techniques in the adequate monitoring and analytical instrumentation to investigate the fate and the behavior of emerging pollutants in wastewater treatment plants and receiving waters. Several recent reviews providing the detailed insight on the analytics of a specific group of micropollutants are given in Table 2.1. Also, Susan D. Richardson published a biennial review covering the recent developments in water analysis for all classes of emerging environmental contaminants over the period of 2007–2008 taking into account 250 most significant references.

2.2 THEORETICAL APPROACHES TO THE ANALYTICS OF MICROPOLLUTANTS

Analytical chemistry of micropollutans usually includes samples that are far from simple, containing numerous components to be analyzed simultaneously or a few target analytes in the presence of many chemical interferents. In such cases, theoretical approaches, e.g., computational chemistry methods and mathematical tools coupled with sophisticated instrumentation are used to assist the analysis.

2.2.1 Computational methods to evaluate the degradation of micropollutants

Computational chemistry is a branch of chemistry that uses principles of computer science to assist in solving chemical problems. The results of theoretical chemistry, incorporated into the efficient computer programs are utilized to calculate the structures and properties of molecules and solids (e.g., the expected positions of the constituent atoms, absolute and relative (interaction) energies, electronic charge distributions, etc.). Computational chemists often attempt to solve the non-relativistic Schrödinger equation, with relativistic corrections:

$$\hat{H}\Psi = E\Psi \tag{2.1}$$

where $\hat{H}$ is the Hamiltonian operator, Ψ is a wave function, and E is the energy. Solutions to Schrödinger's equation are able to describe atomic and subatomic systems, atoms and electrons.

Application of computational chemical methods allows rapid progress in estimating reaction pathways or products from theoretically obtained information on the reaction position or bond conditions. Recent advances in molecular modeling can predict the fate of a huge variety of micropollutants in the presence of different oxidizing agents. Efficient improvement in a predictive molecular-level modeling methods can also enhance the understanding of the fate of new compounds in the water environment.

Computational chemistry methods range from highly accurate to a very approximate methods. *Ab initio* (lat. "from the beginning") methods are directly derived from theoretical principles without using experimental data. Other methods are called empirical or semi-empirical because they include experimental results obtained from the acceptable models of atoms or related molecules, and can approximate or omit various selected elements of the underlying theory. However, to avoid potential errors that occur when one or some elements are omitted, the method is parameterized.

In the recent years, the impact of Density Functional Theory (DFT), one of a recently developed *ab-initio* methods in quantum chemistry, has increased enormously due to less computational requirements however providing the same accuracy of calculations as other computationally intensive methods. DFT is based on the Hohenberg–Kohn theorem that uses electron density instead of the more complex wave function to determine all atomic and molecular properties.

Several examples of computational calculations used to study micropollutants degradation are listed in Table 2.2.

2.2.1.1 Frontier electron density analysis (Frontier orbital theory)

The determination of specific sites of interaction between the two chemical species is of fundamental importance to establish the mechanism of the reaction and also to design the desired products. A number of DFT-based reactivity descriptors, such as the electronic chemical potential, hardness, softness and Fukui function have been derived to determine the specific sites of interaction between the pollutant and an oxidant (e.g., hydroxyl radical). These parameters are associated with the response of the electron density of a system to a change in number of electrons (N) or external potential [v(r)].

Table 2.2 Computational methods used to study advanced (catalytic) oxidation of micropollutants

Compound	AOP	Calculation method (basis set/level)	Software	Identification of products	Ref
Alizarin red	TiO_2 photocatalysis	single determinant Hartree–Fock level with optimization on AM1 level	MOPAC 6.0	GC MS	(Liu *et al.*, 2000)
PAHs in ethanol	Fenton process	PM3	MOPAC 6.0	GC MS with HPLC-UV	(Lee *et al.*, 2001)
17 β-estradiol (E2)	TiO_2 photocatalysis	STO-3G/unrestricted Hartree-Fock (UHF)	Gaussian 98	HPLC +NIST	(Ohko *et al.*, 2002)
Pyridaben	TiO_2 photocatalysis	PM3	Hyperchem, version 5.0.	GC MS	(Zhu *et al.*, 2004)
Polychlorinated dibenzo-p-dioxins (PCDDs),	$Fe(II)/H_2O_2$/ UV	AM1	MOPAC 6.0	GC MS	(Katsumata *et al.*, 2006)
Imazapyr	TiO_2 photocatalysis	NR	MOPAC 6.0	HPLC-UV, LC/ESI-MS	(Carrier *et al.*, 2006)
Alanine	chlorination	AM2	MOPAC 6.0	HPLG,GC-ESI-MS-SCAN (NIST 147)	(Chu *et al.*, 2009)
Levoflaxine	ozone, peroxone	B3LYP/6–31 +G(d,p),BMK/B3LYP	GAUSSIAN 03	GC MS	(De Witte *et al.*, 2009b)
Dinitronaphthalenes	TiO_2 photocatalysis	B3LYP/6–31G*	GAUSSIAN 03	GC MS	(Bekbolet *et al.*, 2009)

Fukui function $f(r)$ is the most important local DFT descriptor. Compared to the frontier orbitals HOMO (highest occupied molecular orbital) and LUMO (lowest unoccupied molecular orbital), the Fukui function contains more detailed information, taking orbital relaxation effects into account (De Witte *et al.,* 2009a). It is based on the Frontier Molecular Orbital (FMO) theory and can be defined as the mixed second derivative of the energy of the molecule with respect to N and $[v(\mathrm{r})]$. Physically, Fukui function reflects how sensitive the chemical potential of the system is to an external perturbation at a particular point. It shows the change in the electron density driven by a change in the number of electrons in its frontier valence region. The Fukui function is the reactivity index for orbital-controlled reactions, where larger Fukui function indicates higher reactivity. Fukui functions per atom i in a molecule can be defined as following:

$$f_i^- = [q_i(N) - q_i(N-1)] \tag{2.2}$$

$$f_i^o = \frac{[f_i^+ + f_i^-]}{2} \tag{2.3}$$

where q_i is the electron population of atom i in the molecule. f_i^- is used when the system undergoes an electrophilic, whereas $f_O\ i$ is valid when the system undergoes a radical attack.

More detailed information about the computational methods and reactivity based descriptors can be found in some recently published books (Parr and Weitao, 1994; Young, 2001).

2.2.2 Chemometrics in analysis

Critical aspect in quantitative analysis is the occurrence of matrix effects (i.e., suppression or, less frequently, enhancement of the analyte signal), which may lead to a significant difference in the response of an analyte in a sample as compared to a pure standard solution. The nature and the amount of co-eluting matrix compounds may vary between the samples in such a way that matrix effects in a series of samples can be difficult to predict. Despite a considerable development in various methods, a challenge in applying sophisticated instrumentation and mathematical tools to develop analytical figure of merits to detect and quantify the target analytes at trace levels remains unsolved. Spectroscopic data can be mathematically modeled by using two different approaches: i) by using hard-modelling-based methods, which require a reaction model to be postulated, or ii) by using soft modelling- based methods, which do not require to know the kinetic model linked to the reactions studied (Escandar *et al.,* 2007).

The standard methods for the second-order approaches to data analysis are parallel factor analysis (PARAFAC), and multivariate curve resolution alternating least squares (MCR), as well as the most recently developed bilinear least-squares (BLLS), unfolded partial least squares/residual bilinearization (U-PLS/RBL), and artificial neural networks followed by the residual bilinearization (ANN/RBL) (Galera *et al.,* 2007). These methods not only simplify the data by reducing dimensionality but also provide visual representation of data as well.

2.2.2.1 Parafac

Multi-way data are characterized by several sets of categorical variables that are measured in a crossed fashion (e.g., any kind of spectrum measured chromatographically for several samples). Determination of such variables would give rise to three-way data; i.e., data that can be arranged in a cube instead of a matrix as in currently available standard multivariate data sets. Assuming trilinearity, PARAFAC decomposes a three-way array X, obtained by joining second-order data from the calibration and test samples:

$$\underline{X} = \sum_{n+1}^{N} a_n \otimes b_n \otimes c_n + \underline{E} \tag{2.4}$$

where $\otimes$ indicates the Kronecker product, N is the total number of responsive components and E is an appropriately dimensioned residual error term. The column vectors a_n, b_n and c_n are usually collected into the score matrix A and the loading matrices B and C.

2.2.2.2 MCR

The goal of multivariate curve resolution (MCR) is to mathematically decompose an instrumental response for a mixture into the pure contributions of each component involved in the system studied. This method is capable of dealing with data sets deviating from trilinearity. Data from the spectroscopic monitoring of a chemical process can be arranged in a data matrix D (r$\times$c), the r rows of which are the number of spectra recorded throughout the process and the c columns of which are the instrumental responses measured at each wavelength (Garrido *et al.,* 2008). The MCR decomposition of matrix D is carried out as following:

$$D = CS^T + E \tag{2.5}$$

where C $(r \times n)$ is the matrix describing how the contribution of the spectroscopically active species n involved in the process changes in the different r rows of the data matrix (concentration profiles). This equation can be solved by either non-iterative methods (e.g., window factor analysis (WFA), sub-window factor analysis (SFA), heuristic evolving latent projections (HELP), orthogonal projection resolution (OPR), parallel vector analysis (PVA)) or iterative approaches (iterative target transformation factor analysis (ITTFA), resolving factor analysis (RFA), multivariate curve resolution-alternating least squares (MCR-ALS). Iterative methods are more frequently used because of their flexibility (assumptions of a model are not required), and ability to handle different kinds of data structures and chemical problems. Moreover, they are able to integrate external information into the resolution process.

2.2.2.3 BLLS

The classic method of least squares is applied to approximate solutions of over determined systems, i.e., systems of equations in which there are more equations than unknowns. This method is usually performed in statistical contexts, particularly regression analysis. Least squares can be interpreted as a method of fitting data.

BLLS starts with a calibration step using the calibration set, in which approximations to pure analyte matrices Sn at unit concentration are found by direct least squares. To estimate the pure analyte matrices Sn, the calibration data matrices $X_{c;i}$ are first vectorized and grouped into a JK × I matrix VX. Then, a procedure analogous to classical least-squares is performed:

$$V_S = V_X \times Y^{T+} \tag{2.6}$$

where Y is an I · N_c matrix collecting the nominal calibration concentrations, N_c is the number of calibrated analytes, and VS (size JK · N_c) contains the vectorized Sn matrices.

2.2.2.4 U-PLS

In the U-PLS method, concentration of the target contaminant is first used in the calibration step (without including data for the unknown sample). The I calibration data matrices Xc i are vectorized and a usual U-PLS model is calibrated with these data and the vector of calibration concentrations. This provides a set of loadings P and weight loadings W (both of size JK × A, where A is the number of latent factors), as well as the regression coefficients v (size A × 1). The parameter A can be selected by techniques such as leave-one-out cross-validation (Galera *et al.*, 2007). If no unexpected interferences occur in the test sample, v can be employed to estimate the analyte concentration:

$$y_u = t_u^T \times \mathrm{v} \tag{2.7}$$

where t_u is the test sample score, obtained by the projection of the (unfolded) data for the test sample X_u onto the space of the A latent factors.

2.2.2.5 ANN

ANN is a computational model simulating the structure and/or functional aspects of biological neural networks. It consists of an interconnected group of artificial neurons and processes information using a connectionist approach to computation. In most cases the ANN is an adaptive system that changes its structure based on external or internal information that flows through the network during the learning phase. In general, neural networks are non-linear statistical data modeling tools. They are utilized to model complex relationships between the inputs and outputs or to find patterns in available data. Typically, an ANN comprises three layers: input, hidden, and output. Usually, the number of input neurons equals A, where, A principal components (PCs) $\ll$ JK (the number of channels in each dimension). The value of A is estimated by computing the % of variance explained by the PCs of the unfolded training data matrix (size I×JK, I is the number of training samples), and selecting the first A PCs which explain more than a certain % (i.e., 99%) of the total variance (Galera *et al.*, 2007).

The training/prediction scheme works properly for the test samples, which have a composition that is representative of the training set. However, when unexpected constituents occur in the test sample, its scores will not be suitable for the analyte prediction using the currently trained ANN. In this case, it is necessary to resort to a technique that marks the new sample as an outlier, indicating that further actions are necessary before ANN prediction, and then isolates the contribution of the unexpected component from that of the calibrated analytes, in order to recalculate appropriate scores for the test sample (Galera *et al.*, 2007).

2.3 INSTRUMENTAL METHODS

The authentic characteristics of micropollutants, such as their occurrence at trace concentration levels, the presence of extremely diverse groups, and etc. make procedures of the detection and analysis quite challenging. Traditionally, structure proof involves several steps: purification, functional group identification, and establishment of atom and group connectivity. The modern instrumental methods have the ability to run reactions, purify products, and determine structures on the nanogram scales with the subsequent increase in the rate at which structural information can be obtained. This has resulted in an exponential growth of chemical knowledge and is directly responsible for the explosion of information being continually published in the chemical literature.

2.3.1 Sample preparation

Many techniques for separating and concentrating the species of interest have been devised in the past several decades. Such techniques are aimed at exploiting differences in physico-chemical properties of various components present in a mixture and can be divided into extraction and chromatographic separation methods.

2.3.1.1 Sample extraction

Generally, the pretreatment or extraction step plays an important role in determining the overall level of analytical performances in practice. The development of new methods for sample preparation, which can dramatically reduce the required amounts of the reagent and organic solvents, also improves the characteristics of other methodologies, which cannot be directly applied to the samples as electrochemical and chromatographic methods can (Armenta *et al.,* 2008). Current trends in sample handling focus on the development of faster, safer, and more environment friendly extraction techniques (Rubio and Perez-Bendito, 2009; Tobiszewski *et al.,* 2009). In order to select the most appropriate and efficient extraction method, several points should be taken into account: i) the solubility of the compound and ii) the difference between concentrations of the parent compound and its degradation products. Moreover, degradation products to be separated by, for instance, gas chromatography (GC) are generally dissolved in organic solvent instead of water, thus only small volumes can be injected to obtain the results. Therefore, sample concentration procedures, which provide a solvent change are needed (De Witte *et al.,* 2009a).

Solid-phase extraction (SPE) is one of the leading techniques for the extraction of pollutants from the aquatic systems (Kostopoulou and Nikolaou, 2008). One of the most important parameters in the application of a SPE method is the selection of an appropriate solid sorbent to the target analyte as well as the use of solvents for washing and the subsequent elution of the target contaminant. The sorbent can be packed into small tubes or cartridges, such as a small liquid chromatographic column or it is also available in the shape of discs with a filtration apparatus (Chang *et al.,* 2009). No losses of the reaction intermediates should occur during sample loading while 100% recovery is desired in the elution step (De Witte *et al.,* 2009a). Due to the wide range of polarities of micropollutants, non-selective sorbents such as silica (C18) or polymer-based resins are most commonly used. Recent developments in this field are mainly related to the use of new sorbent materials such as molecularly imprinted polymers (MIPs). Depending on the type of cartridges, the recovery rates of an identical analyte may vary from 10 to 90%. Moreover, Gatidou and co-workers (2007) compared several types of cartridges for the extraction of nonylphenol (NP), nonylphenol monoethoxylate (NPIEO), triclosan (TCS), bisphenol A (BPA), and nonylphenol diethoxylate (NP2EO) from aqueous solution. The analytical results of the target compound were compared for different types of the SPE cartridge: C18, EnviChrom-P, Isolute NV+, and Oasis HLB. Among them, only C18 demonstrated sufficient recoveries for the majority of the compounds, whereas the rest of studied cartridges were selectively efficient (Gatidou *et al.,* 2007).

Nowadays, enhanced solvent-extraction techniques are widely accepted for the extraction of pollutants from environmental solid matrixes, however, serious limitations that relates to the cleanup step still remains unsolved. The latest trends for the purification of collected extracts involve the use of some fast and/or simplified procedures to treat the liquid samples, e.g., solid-phase (SPME) and liquid phase (LPME) microextraction. In the past few years, matrix solid-phase dispersion (MSPD) became very popular in the extraction of organic pollutants from a variety of solid environmental samples and headspace microextraction procedures (e.g., HS-SPME) and gradually developed into the preferred tools for the extraction of volatile and semivolatile compounds (Rubio and Perez-Bendito, 2009). Alternative approaches to the sample preparation process are extraction techniques based on exploitation of microwave energy (i.e., microwave-assisted extraction (MAE)) and micelle creation (i.e., micelle-mediated extraction (MME), where cloud-point extraction (CPE) is of special importance. In literature, MAE is usually reserved for the liquid –liquid (LLE) and solid–liquid (SLE) extraction, which are assisted by the microwave energy (Madej, 2009).

The SPME method primarily involves the adsorption of an analyte onto the surface of a sorbent-coated silica fiber. The advantage of SPME is the direct injection of the adsorbed analyte into the chromatograph. The sensitivity of the SPME method can be assured, since the polymeric stationary phases used in SPME have a high affinity for organic molecules (Chang *et al.,* 2009). The practical application of SPME has various aspects. Direct and headspace SPME can be differentiated, depending on where the coated fiber is placed for the extraction. Occasionally, a vigorously selected derivatization reagent (e.g., BSA, BSTFA, MSTFA, TMCS, HMDS) can be added to enhance the performance of the selected analytical method. In addition, operational factors, such as sample stirring speeds as well as the extraction time and temperature required to reach the equilibrium between the aqueous and stationary phases on the SPME fiber should be adjusted. If all operational parameters are fixed, the analysis can be automated with the reduced usage of solvents. Moreover, the fiber can be reused several times and recycled with a proper care (Chang *et al.,* 2009).

2.3.1.2 Chromatographic separation

The selection of the chromatographic separation technique is based on the polarity and thermal stability of the target compound. The most performed indispensable chromatography separation types are GC (gas chromatography) and LC (liquid chromatography). Recent advances in separation techniques of special relevance to environmental analysis of micropollutants and their degradation products have been promoted by the use of thin layer chromatography (TLC), ultraperformance liquid chromatography (UPLC), hydrophilic interaction chromatography (HILIC), fast gas chromatography (FGC), comprehensive two-dimensional gas chromatography (GC × GC), and capillary electrophoresis-based microchips (μCE) (Rubio and Perez-Bendito, 2009). LC is probably the most versatile separation method, as it allows separation of compounds of a wide range of polarity with little effort, compared to GC-MS, in sample preparation. Semi-polar compounds can be separated using reverse-phase columns, while hydrophilic columns can facilitate the quantification of target polar compounds (Chang *et al.,* 2009). The fastest growing chromatography trend still remains in the use of ultra-performance liquid chromatography (UPLC). UPLC is a recently developed LC technique that uses small diameter particles (typically 1.7μm) in the stationary phase and short columns, which allow higher pressures and, ultimately, narrower LC peaks (5–10 s wide) (Richardson 2009). UPLC has such advantages as providing narrow peaks, improved chromatographic separations, and extremely short analysis times, often to 10 min or less.

2.3.1.3 Capillary electrophoresis (CE)

The physical mechanisms of capillary electrophoresis (CE) are similar to LC. It is also a water based separation technique, which is well suited to the analysis of inherently hydrophilic compounds. The main difference from LC, which is based on the partition of the solutes between the mobile phase and the stationary phase, is that CE is derived from the differences in charge-to-mass ratio of the molecules. Therefore, a totally different selectivity is expected for the analytes, thus providing a separation method complementary to LC (Pico *et al.,* 2003).

All three separation techniques including their advantages and disadvantages are summarized in Table 2.3.

Table 2.3 Comparative analysis of separation techniques (Adapted from Pico *et al.* (2003)

Technique	Advantages	Disadvantages	Solutions
GC	High resolving power and ability to resolve individual analytes	Inadequate for polar, thermo-labile and low-volatility compounds	Derivatisation
	High sensitivity and selectivity Existence of mass-spectrum libraries for screening unknown samples	High consumption of expensive, high-purity gases	
LC	Application to virtually any organic solute regardless of its volatility or thermal stability	Insufficient separation efficiency and selectivity	Development of more efficient and selective columns materials (immuno-sorbents, MIPs and restricted access materials)
	Both mobile and stationary phase compositions are variables	Large amounts of expensive and toxic organic solvent used as mobile phase	
	Capable of automation and miniaturisation (microchip technology)		
CE	High separation efficiency	Inadequate limits of detection Lack of selective detectors	Sample enrichment (SPE, stacking)
	Small consumption of expensive reagents and toxic solvents		Increase detection path-length
	Capable of automation and miniaturisation (microchip technology)		Development of coupling methods to combine CE with highly selective detectors

2.3.2 Detection of micropollutants transformation products

Identification of micropollutants transformation products is a challenge that resides in obtaining high-quality data suitable for the detection from the available analytical technologies. These analytical technologies (liquid chromatography (LC), mass spectrometry (MS), fragmentation pattern analysis (MS2), ultraviolet/visible spectroscopy (UV/Vis), nuclear magnetic resonance (NMR) and experimental validation by standard compounds) are summarized in Figure 2.1. In addition, information already available in literature and databases (NIST, Wiley, etc.) would also be an important identification tool to determine the transformation products of target micropollutans.

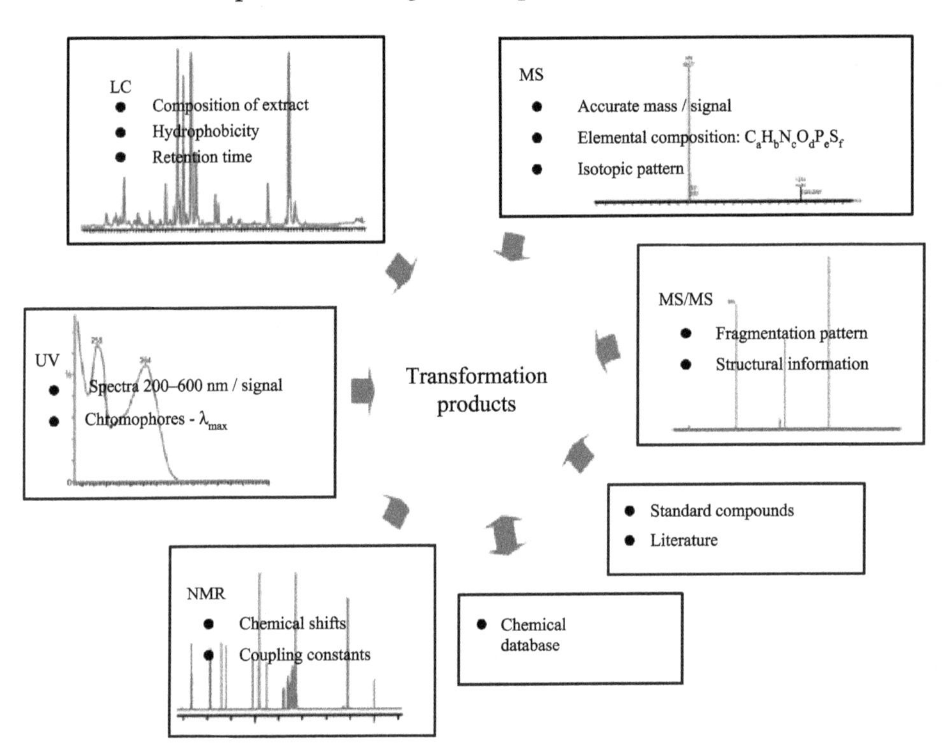

Figure 2.1 The most common methods to identify the transformation products (Adapted from Moco *et al.* (2007)

However, identification of transformation products is quite a challenging task due to several problems, associated with the identification of unknown structures. Usually, they present in low concentrations in the samples and standard material for structure elucidation is seldom available (De Witte *et al.*, 2009a). The scheme of comprehensive structure elucidation of micropollutants transformation products can include several detection methods (e.g., HPLC/NMR, UV/MS, etc.) (Kormos *et al.,* 2009).

2.3.2.1 Mass spectrometry

Currently, the most prevailing approach designed to analyze emerging pollutants incorporates a mass-based analysis process. Generally speaking, the mass-based methods employing mass spectrometry (MS) show relatively low detection limits as compared to other methods. MS is a spectrometric method that allows the detection of mass-to-charge species that can point to the molecular mass (MM) of the detected compound.

Mass spectrometry involves absorption of energy at particular frequencies. During the MS determinations, a molecule is purposely broken into pieces, these pieces are identified by the mass, and subsequently the original structure is then inferred from these pieces. Mass spectrometers can be divided into three fundamental parts: i) the ionization source, ii) the analyzer, and iii) the detector (Scheme 2.1). Different mass analyzers have different features, including the m/z range that can be covered, the mass accuracy, and the achievable resolution as well as the compatibility of different analyzers with different ionization methods. GC-MS instruments make use of the hard-ionization method, electron-impact (EI) ionization, while LC-MS mostly uses soft-ionization sources (e.g., atmospheric pressure ionization (API) (e.g., electrospray ionization (ESI)) and atmospheric pressure chemical ionization (APCI)) (Moco *et al.,* 2007). In LC-MS applications, the mass detection of a molecule by MS, is conditioned by the capacity of the analyte to ionize while being part of a complex mixture and, therefore, micropollutants unable to ionize, cannot be detected.

2.3.2.1.1 Recent advances in MS based techniques. Various configurations of mass spectrometers used for MS applications, in terms of ion acceleration and mass detection, ion-production interfaces and ion fragmentation capabilities are currently available. Liquid and gas chromatography, in combination with mass spectrometry (LC-MS, GC/MS) and tandem MS (LC-MS2, GC/MS^2), continue to be the predominant techniques for the identification and quantification of organic pollutants, and their transformation products, in the environment. New methods reported include the utilization of UPLC/MS/MS, LC/TOF-MS with accurate mass determination, and online-SPE/LC/MS^2. The recent developments in UPLC allowed to vastly improve chromatographic resolution (as compared to the conventional LC), run times (often less than 10 min), and to minimize the matrix effects. Recent advances in capillary electrophoresis-mass spectrometry (CE-MS) have rendered this technique more competitive in the analysis of environmental samples; however, the inherently small (trace) amounts of pollutants present in the environment, and the high complexity of environmental sample matrixes, have placed strong demands on the detection capabilities of CE (Rubio and Perez-Bendito, 2009).

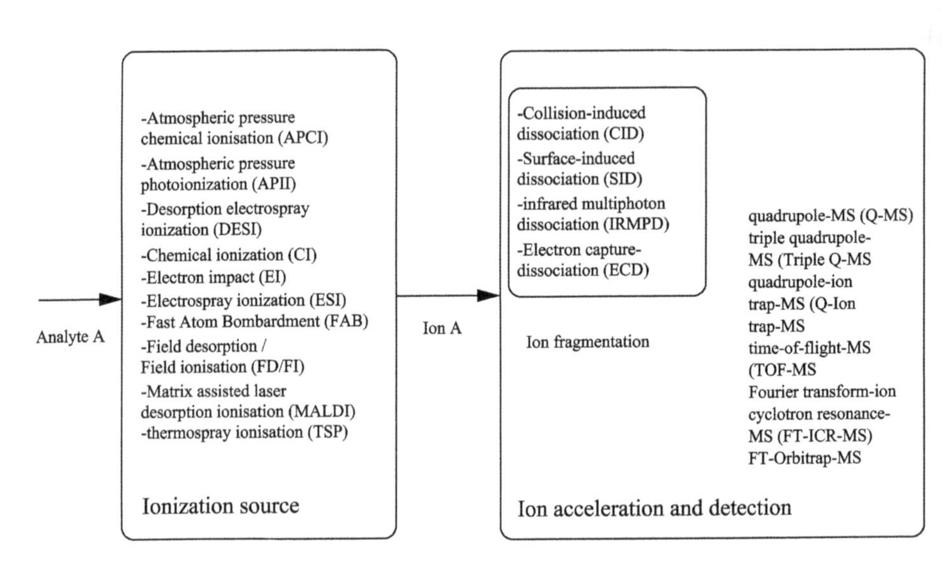

Scheme 2.1 Simplified configuration of MS (Adapted from Moco *et al.* (2007)

Also, there is a tremendous increase in the use of time-of-flight (TOF)-mass spectrometry (MS) and quadruple (Q)-TOF-MS for the structural elucidation and compound confirmation. As a rule, (Q)-MS instruments have a mass resolving power that is 4 times less than that of a time-of-flight (TOF)-MS, while a Fourier transform (FT)-ion cyclotron (ICR)-MS can reach a resolving power of higher than 1,000,000 (i.e., 400 times greater than a Q-MS) (Richardson, 2009). The use of hybrid QTOF, instead of single TOF, offers more possibilities in screening and identification of target micropollutants, so it is feasible to access the MSE acquisition mode, where low and high collision-energy full-scan acquisitions can be performed simultaneously, resulting in valuable fragmentation information for use in elucidation and confirmation of the unknown compounds (Ibanez *et al.*, 2008). TOF-MS and Q-TOF-MS provide an increased resolution capability (typically 10 000–12 000 resolution), which allows the precise empirical formula assignments for the unknowns and also provides extra confidence for positive identifications in the quantitative work (Richardson, 2009). Especially, TOF-MS and Q-TOF-MS are effective means in the detection of pharmaceuticals, endocrine disrupting compounds (EDCs), and pesticide degradation products.

A higher mass accuracy facilitates a finer distinction between closely related mass-to-charge signals, so the quality and the quantity of assignments of mass signals to metabolites can be much improved by using high-resolution and ultra-high-resolution accurate mass spectrometers (Moco *et al.,* 2007).

In addition, liquid chromatography (LC)/electrospray ionization (ESI)- and atmospheric pressure chemical ionization (APCI)-MS methods continue to dominate the new methods developed for the detection of emerging contaminants, and the use of multiple reaction monitoring (MRM) with MS/MS has become commonplace for the quantitative environmental analysis. Also, atmospheric pressure photoionization (APPI) is used with LC/MS to provide the improved ionization for more non-polar compounds, such as polybrominated diphenyl ethers (PBDEs). Furthermore, FT-MS instruments, such as the cyclotron (FT-ICR-MS) and the Orbitrap type (FT-Orbitrap-MS), enable measurements at a higher mass accuracy in a wider dynamic range. FT-ICR-MS has the highest mass-resolving power so far reported for any mass spectrometer ($>$1,000,000) and a mass accuracy is generally within 1 ppm (Hogenboom *et al.,* 2009). The recently developed FT-Orbitrap-MS has a relatively modest performance compared to the FT-ICR-MS (maximum resolving power $>$100,000 and 2 ppm of mass accuracy with an internal standard), but it is a high-speed, high ion-transmission instrument, due to shorter accumulation times (Moco *et al.,* 2007). Hyphenation of UPLC and TOF-MS is an efficient, advanced approach for rapid screening of non-target organic micropollutants in water. Therefore, UPLC provides a fast chromatographic run with improved resolution, minimizing the co-elution of components. Then again, the huge amount of information provided by TOF-MS, together with the measurement of accurate mass for the most representative ions, facilitates confident identification of non-target compounds in samples (Ibanez *et al.,* 2008).

In general, the chromatographic parameters (such as temperature, pH, column, flow rate, eluents, gradient), injection parameters, sample properties, MS and MS^2 parameters (calibration and instrumental parameters, such as capillary voltage and lens orientation), and the remaining parameters related to the configuration of the system can affect the performance of analyses. An adequate configuration should be adopted, taking into account the aim of the analyses and the current limitations of the instruments.

2.3.2.1.2 Data interpretation. High resolution instruments provide very high mass accuracies, and the range of possibilities for chemical formula is limited, especially for lower m/z values and therefore, it is more proficient to determine the correct molecular formula than low resolution instruments. However, the number of possible chemical structures increases with increasing molecular mass values.

Important aspect that must be taken into account when determining the chemical structure of the target compound is the algorithm used for the calculation. There are more possible mathematical combinations of elements that fit certain molecular masses than the number of chemical formulas that exist chemically. The chemical rules (e.g., the octet rule) dictate certain limitations on chemical bonding derived from the electronic distribution of the participating atoms present in molecules (e.g., nitrogen rule). Also, the number of rings and double bonds can be calculated from the number of C, H atoms in a molecule.

The practical approach to narrow the choice of chemical structures is the isotope labeling. Isotopically labeled standards (deuterated or ^{13}C, ^{15}N-labeled) allow a more accurate quantification of a target compounds in a variety of sample matrixes. This is especially relevant for wastewater and biological samples, where matrix effects can be substantial (Richardson, 2008). For the most small organic molecules, the intensity of the second isotopic signal, corresponding to the ^{13}C signal, can indicate the number of carbons that the molecular ion contains (natural abundance of ^{13}C is 1.11%) (Moco *et al.*, 2007). Kind and Fiehn (2006) stated that this strategy can remove more than 95% of false positives and can even outperform an analysis of accurate mass alone using a (as yet non-existent) mass spectrometer capable of 0.1 ppm mass accuracy. Several works of isotope labeling in the analysis of transformation products have been reported (Vogna *et al.*, 2002; McDowell *et al.*, 2005).

Another possibility to accurately determine a target compound is isolating one ion and performing MS^2 to the successively obtained fragments. It is extremely informative for tracking functional groups and connectivity of fragments for elucidating the structures of degradation products.

When a separation method is coupled to the mass spectrometer, retention time is a parameter that can give information about the polarity of the metabolite. Nowadays, retention-time variation is relatively low in stabilized (LC or GC)-MS setups, thus allowing a direct comparison of chromatograms and the construction of databases (Moco *et al.*, 2007).

The ability to assign the reaction intermediates using MS resides in the possibility of combining different features of the MS analysis (accurate mass, fragmentation pattern, and isotopic pattern) with additional experimental parameters (e.g., retention time) and its confirmation with standard compounds. Experimental spectra can be compared against a home-made mass-spectral library (either empirical or theoretical), hence automating an efficient screening for many different compounds. Home made libraries need to contain a large number of contaminants, and exact masses need to be included in the databases for the correct candidate assignation. When a match is unsatisfactory, the deconvoluted MS spectra can be used to investigate the identity of unknown compounds not present in the library (Ibanez *et al.*, 2008).

2.3.3 UV-Visible spectroscopy

UV-Vis spectroscopy was one of the earliest techniques to be combined with other techniques when chromatographic separation was incorporated with UV-Vis detection. Although it is common to use fixed wavelength detection, rapid digital scanning can permit essentially continuous spectral scanning of chromatographic effluents.

Absorption of radiation in the visible and ultraviolet regions of the electromagnetic spectrum results in electronic transitions between molecular orbitals. The energy changes are relatively large, corresponding to about 105 J mol^{-1}, which corresponds to a wavelength range of 200–800 nm or a wave number range of 12 000–50 000 cm^{-1} (Fifield and Kealey, 2000). The use of UV-Vis for the determination of the degradation products usually includes the measurement of absorbance spectra of the sample or individual degradation product after the chromatographic separation. The appearance and disappearance of the absorbance maxima is directly related to the presence of the particular functional groups. Unsaturated groups, known as chromophores, are responsible for the absorption mainly in the near UV and visible regions and are of most value for the identification purposes and for quantitative analysis. The positions and intensities of the absorption bands are sensitive to the substituents, which are close to the chromophore, as well as to conjugation with other chromophores, and to the solvent effects. UV-Vis generally requires quite low concentrations of the analyte (absorbances less than 2). However, the presence of the same chromophore in different compounds will result in the same UV-Vis spectra. Therefore, despite the use of UV-Vis is useful for the estimation of the degradation products, more advanced analytical methods are needed in order to identify possible reaction intermediates (De Witte *et al.*, 2009a).

New developments in this technique include commercial versions of UV-vis imaging systems, attenuated total reflection (ATR) spectroscopy, FT-UV spectroscopy optimization, and various applications of fiber optic based and high-resolution techniques (Richardson, 2009).

2.3.4 NMR spectroscopy

Nuclear magnetic resonance (NMR) spectroscopy has become an important technique for the determination of chemical and physical properties of a wide array of organic compounds. Such determination is possible due to the absorption of energy at particular frequencies by atomic nuclei when they are placed in a magnetic field. Nuclei, which have finite magnetic moments and spin quantum numbers of I = ½ such as ^{1}H, ^{11}B, ^{13}C, ^{15}N, ^{17}O, ^{19}F, and ^{31}P are the most useful and common in NMR measurements. Since hydrogen and carbon are the most common nuclei found in organic compounds, the ability to probe these nuclei by NMR is invaluable for the organic structure determination. Unfortunately, as the nuclear transition energy is much lower (typically of the order of 104) than the electronic transition, NMR is relatively un-sensitive in comparison to other techniques, such as UV/Vis spectroscopy (Wishart, 2008). Furthermore, the signal-to-noise ratio (S/N) in NMR depends on many parameters (e.g., magnetic field strength of the instrument (B_0), concentration of the sample, acquisition time (NS), and the measurement temperature).

However, NMR is perhaps the most selective analytical technique currently available, being able to provide unambiguous information about a target molecule. NMR can elucidate chemical structures, and can provide highly specific evidence for the identification of a molecule (Moco *et al.,* 2007). Moreover, NMR is a quantitative technique, as the number of nuclear spins is directly related to the intensity of the signal.

NMR can be directly applied for the structure elucidation. ^{1}H is the most used nucleus for NMR measurements due to its very high natural abundance (99.9816–99.9974%) and good NMR properties. In fact, several NMR techniques are available for a great variety of purposes, e.g., one-dimensional 1H NMR, two-dimensional (2D)-NMR, homonuclear ^{1}H-2D spectra (e.g., COSY, TOCSY and NOESY) and heteronuclear 2D spectra that can be acquired for detecting direct ^{1}H-^{13}C bonds by HMQC or, over a longer range, HMBC. Nowadays, the detection limit in a 14.1 Tesla (600 MHz for 1H NMR) instrument, is in the microgram (^{1}H-^{13}C NMR) or even sub-microgram region (^{1}H NMR). The sensitivity of NMR has been improving over the years, increasing the suitability of this technique for various analytical applications (Cardoza *et al.,* 2003).

However, the direct measurements by NMR face several difficulties: i) the abundance of protons an the reaction media (usually water) prevents the attribution of chemical shifts to the individual proton, and ii) the extraction of the target compound is essential prior to NMR analysis.

There are different configurations for coupling chromatography to NMR (Exarchou *et al.,* 2003; Exarchou *et al.,* 2005). More recently the on-line coupling of LC to SPE and subsequently to NMR became available and improved some of the existing analytical barriers of the previous modes. In this configuration, the chromatographic peaks are trapped in SPE cartridges and can be concentrated up to several times by multi-trapping into the same cartridge (Moco *et al.,* 2007).

Recently, the separation of 2-phenyl-1,4-benzopyrone and phenolic acids present in Greek oregano extract was accomplished by LC-UV-SPE-NMR-MS (Exarchou *et al.*, 2003). The compounds were separated by LC, trapped in SPE cartridges and subsequently eluted for NMR and MS acquisition. The work demonstrated that two related compounds co-eluting in the LC (and therefore trapped into the same cartridge) could be readily distinguished by MS and NMR (Exarchou *et al.*, 2003). This method is applicable for the analysis of rare or/and unknown compounds in complex mixtures, since it allows separation, concentration and NMR acquisition of the analyte within a single system. Moreover, concurrent interpretation of UV, MS, and NMR data obtained for every trapped peak led to the unequivocal assignment of their structure.

The identification of micropollutant degradation products can be aided by various methods, such as MS or NMR, but often the full chemical description of a molecule is only achieved by integrating the compound information taken from different sources (Moco *et al.*, 2007). Advanced analytical instrumentation employ MS to determine the MS fragmentation pathways (i.e., cleaved moieties) whereas ^{1}H and ^{13}C NMR is utilized for structural confirmation (Kormos *et al.*, 2009). Therefore, the combination of MS with NMR is one of the most powerful strategies for the identification of an unknown molecule. To combine the advantages of NMR spectroscopy and MS spectrometry, the extended hyphenation of LC-NMR-MS has been used to identify compounds (Exarchou *et al.*, 2003). The most efficient way to seize the advantages of both technologies, is to use them in parallel or, if possible, on-line. However, due to the complex analytical set-up, it is still most common to undertake analyses by LC-MS and LC-(SPE)-NMR separately (Moco *et al.*, 2007). In summary, high information content of NMR experiments makes this an attractive technique for the analysis of contaminant-transformation processes, especially when coupled with a separation method. Because of the complementary nature of the results provided by each technique, the combination of NMR and MS/MS analysis is an especially powerful approach for elucidating the structure of new transformation products (Cardoza *et al.*, 2003).

2.3.5 Biological assessment of the degradation products

Biological assessment of the micropollutant degradation products complements chemical analysis in terms of information provided on toxicity, estrogenicity and antibacterial activity of the discharged wastewater. The use of biological organisms is vital in the measurement and the evaluation of the potential impact of these contaminants (Wadhia and Thompson, 2007).

2.3.5.1 Ecotoxicological assessment of environmental risk (toxicity)

An environmental toxicant can be defined as a substance that, in a given concentration and chemical form, challenges the organisms (bioindicators) of the ecosystem and causes adverse or toxic effects (Lidman, 2005) The structure of micropollutant molecule will account for its physico-chemical characteristics and properties and subsequently, its toxicity. Molecules with similar structure have similar properties, and thus the potential for the similar toxicity (Hamblen *et al.,* 2003). Therefore, Veith *et al.* (1988) constructed a model using a K-nearest neighbor pattern recognition technique based on eight molecular topological parameters as molecular descriptors. To describe the interrelatedness of chemical structures. K-nearest neighbors explained over 90% of the observed variability in the 8 descriptors (Veith *et al.,* 1988).

Environmental samples can be ecotoxicologically tested using any level of biological organization from molecular to whole organisms and populations, communities or assemblages of organisms. However, the assessment of toxicity will be reliable only if several specific aspects will be considered during the analysis with standardized test protocols: i) environmental samples taken for testing with bioassays need to be representative, and the collection, storage and preparation procedures must not result in change in toxicity of the sample; ii) a measure of test variability needs to be ascertained; and iii) the effect of confounding variables (e.g., pH, dissolved solids, and Eh) also needs to be established.

Various ecotoxicity tests have been developed to characterize the toxicity of individual chemicals. The adoption of these tests for the toxicity evaluation of environmental samples led to contaminated media tests and/or bioassays. A bioassay can be defined as a biological assay performed to measure the effects of a substance on a living organism, so bioassays can provide qualitative or quantitative evaluation of a selected system. In the latter case, the assessment often involves an estimation of the concentration or the potency of a substance by measuring selected biological responses. A toxicity test can be considered a bioassay that allows measurement of damage (Wadhia and Thompson, 2007).

Recently, the benefits of using rapid, sensitive, reproducible and cost-effective bacterial assays such as bioluminescent bacteria (BLB) were acknowledged. One of the most recognized bioluminescent bacteria is *Vibrio fisheri* (NRRL B-11177) that emit light as a result of their normal metabolism (about 10% of its metabolic energy), and the intensity of the light emitted is an indication of their metabolic activity (Gu *et al.,* 2002). When these bacteria are exposed to a toxic substance, the light emission is reduced, thus providing a measure of the acute toxicity of a sample. Its advantages, over alternatives, are simplicity of operation, speed and low cost. Moreover, this is a standardized organism that has been studied very well, and a number of different systems are commercially available (Farre *et al.,* 2007).

The need to bioanalyze a large number of environmental samples in a relatively short period of time led to the development and the increased significance of fast, miniaturized tests for toxicity, so-called microbiotests (also known as alternative tests and second-generation tests) (Wolska *et al.,* 2007). Microbiotests have many undoubted advantages, such as: i) relatively low cost per analysis, ii) operation with small sample volumes, iii) no requirement to culture test organisms (organisms are stored in cryptobiotic form, i.e., rotifers as cysts, crustaceans as resting eggs, and algae as cells immobilized on specific medium), iv) possibility of working with several samples at once, v) short response time, vi) repeatability and reproducibility of data, and vii) tests can be performed under laboratory conditions. Furthermore, the microbiotests can be used by personnel without any special training and previous experience of working with bioindicators. Microbiotests are also widely utilized in commercially available systems such as ToxAlert 10 and ToxAlert 100 (Merck); LUMIStox (Dr. Bruno Lange); ToxTracer (Skalar); Biotox and The BioToxTM Flash, Toxkit (Aboatox); Microtox and microtox SOLO, DeltaTox Analyzer (AZUR Environmental); ToxScreen (CheckLight).

2.3.5.2 Assessment of estrogenic activity

Routledge and Sumpter (1997) anticipated that both, the position and branching of the substituent of alkylsubstituted phenolic compounds affected their estrogenic potency. In particular, optimal activity was associated with a tertiary-branched alkyl moiety of 6–8 carbon atoms substituted in the 4-position to the hydroxy group (Routledge and Sumpter, 1997). Thus, theoretically, the estrogenic activity of every compound is related to its chemical structure, e.g., relative position of the hydroxyl phenolic group in the ring due to its high affinity with the estrogen receptor, which is responsible for its estrogenic potential (Streck, 2009).

Biological analysis methods vastly improved over the last decade, especially with advances in the development of bioassays for the estrogenicity assessment. Bioassays (e.g., YES, MELN tests) have the advantage to measure the estrogenic effect related to hormones and other estrogenic disruptors present in the samples, so they can be better adapted to screen estrogenic disruption in aquatic environments exposed to urban and industrial sources of contamination. However, the possible inhibition effect from a mixture of pollutants needs to be taken into account by performing chromatographic fractionation of samples and biological testing of the isolated fractions individually (estrogenic potency). The tests of estrogenicity assessment are summarized in Table 2.4.

Table 2.4 Tests for assessment of the estrogenic activity in transformation products of estrogenic compounds

Bioassay	Basics of the method	Reference
The yeast estrogen screen (YES) test	recombinant yeast cultures expressing human estrogen receptor	(Nelson *et al.*, 2007) (Salste *et al.*, 2007)
E-screen test	proliferation of human breast-cancer cells MCF-7 under estrogenic control	(Korner *et al.*, 2000)
ER-CALUX	human breast-cancer cells T47D stably transfected with luciferase reporter gene	(Murk *et al.*, 2002)
MELN	human breast-cancer cells MCF-7 stably transfected with luciferase reporter gene	(Pillon *et al.*, 2005)

2.3.5.3 Assessment of antimicrobial activity

There are many concerns about the widespread distribution of antimicrobial agents and preservatives (e.g., triclozan,) in the environment, which lead to the development of cross-resistance to antibiotics, the adverse effects on ecological health, and the formation of more toxic pollutants under different conditions. For instance, triclosan has been incorporated into a broad array of personal care products (e.g., hand disinfecting soaps, medical skin creams, dental products, deodorants, toothpastes), consumer products (e.g., fabrics, plastic kitchenware, sport footwear), and cleaners or disinfectants in hospitals or households thus its transformation may lead to adverse effects on various environmental systems (Lange *et al.*, 2006; Roh *et al.*, 2009).

The classic measure of antimicrobial potency is the minimal inhibitory concentration (MIC) for a particular agent with a given pathogen. For instance, by relating the MIC to concentrations of the selected pathogen in human tissues, it has been possible to develop empirical relationships between exposure, potency and patient response (Drlica, 2001). Reduction of microbial activity can be estimated by the cell growth measurement in aqueous solution or in agar plates. Tests on *Pseudomonas putida*, *Escherichia coli*, etc. usually indicate whether the degradation process of a target contaminant is sufficient for the reduction of its antimicrobial activity.

2.3.5.4 Biosensors

Biosensor can be defined as an integrated receptor-transducer device, which is capable of providing selective quantitative or semi-quantitative analytical information using a biological recognition element. In biosensors, a biological unit (e.g., an enzyme or an antibody), which is typically immobilized on the surface of the transducer (electrode), interacts with the analyte (which contains a target compound) and causes a change in a measurable property within the local environment near the transducer surface, thus converting a (bio)chemical process into a measurable electronic signal. The main factors that influence the biosensor response are the mass-transport kinetics of analytes and products, as well as loading of the sensing molecule. Biosensors are usually categorized by the transduction element (e.g., electrochemical, optical, piezoelectrical or thermal) or the biorecognition principle (e.g., enzymatic, immunoaffinity recognition, whole-cell sensor or DNA). Biosensors offer simplified, sensitive, rapid and reagentless real-time measurements for a wide range of biomedical and industrial applications.

Recently, enzymatic biosensor (*tyrosinase*) was applied for the detection of BPA (Carralero *et al.,* 2007), 17-β-estradiol (Notsu *et al.,* 2002), PAH (Fahnrich *et al.,* 2003) and surfactants (Taranova *et al.,* 2004). Optical biosensors with high sensitivity (up to ng range) were successfully used for the monitoring of pesticides, EDCs, and surfactants (Skladal, 1999; Tschmelak *et al.,* 2006).

Biosensor technology is a rapidly expanding field of research that has been transformed over the past two decades through new discoveries related to the novel material technologies, means of signal transduction, and powerful computer software to control devices (Farre *et al.,* 2009). More detailed information on biosensors is available in Chapter 3.

2.4 IDENTIFICATION LEVELS OF MICROPOLLUTANTS TRANSFORMATION PRODUCTS

Recent developments in analytical methodologies led to the variations in the scientific evidence that greatly support the identification of micropolutants transformation products. Therefore, so called metabolomics approaches include targeted analyses that relates to residue and a target contaminant analysis as well as untargeted analyses in which the goal is to measure as many compounds as possible. In the latter case, identification is not the primary step in data processing. The analytical information obtained from the profiles is transformed to coordinates on the basis of mass, retention and amplitude.

The GC/MS spectra are readily available for many widely studied compounds, however, for the most of micropollutants, especially taking into account the constantly growing number of the potential emerging compounds, such databases do not exist. Thus, De Witte *et al.* (2009a) proposed the utilization of the methodology created by the Chemical Analysis Working Group (CAWG) as a basis for the identification of pharmaceutical degradation products. According to the CAWG methodology, four levels of identification can be distinguished for previously characterized, identified and reported compounds. Level 1 requires a minimum of two independent data compared with a standard compound, analyzed under identical experimental conditions. Level 2 identification is associated with putatively annotated compounds (e.g., without chemical reference standards, based upon physicochemical properties and/or spectral similarity with public/commercial spectral libraries) (Sumner *et al.,* 2007). In case of the characterization based upon characteristic physicochemical properties of a chemical class of compounds, or by spectral similarity to known compounds of a chemical class, the identification level is assigned to level 3. And, finally, level 4 transformation products are labeled as unknown compounds – although unidentified or unclassified these metabolites can still be differentiated and quantified based upon available spectral data.

De Witte *et al.* (2009a) also proposed four approaches as a novel identification scheme based on CAWG identification levels for metabolites in biological organisms and the notations of Doll and Frimmel. Therefore, four approaches based on the subsequent identification method engaged with available analytical tools are depicted in Figure 2.4.

The proposed scheme is applicable not only to the pharmaceutical degradation products but also to the broader range of micropollutants (e.g., pesticides, etc.). Every approach is connected to the relevant level of identification. For instance, analysis of a standard compound in approach A, next to the analysis of the sample or the parent molecule/analogous products in approach B can strengthen identification. Approach C proposes the involvement of MS spectra bases, whereas the approach D involves only the sample analysis. Approach D is also divided into two levels because the difference has to be made between the identification based on the extended MS fragmentation (tentative identification) or/and on the molecular weight or UV spectrum determination (indicative determination). Level 1 and level 3 are not commonly possible, whereas level 2, 4 and 5 represent the classification levels more likely using the current state-of-art technologies (De Witte *et al.,* 2009a).

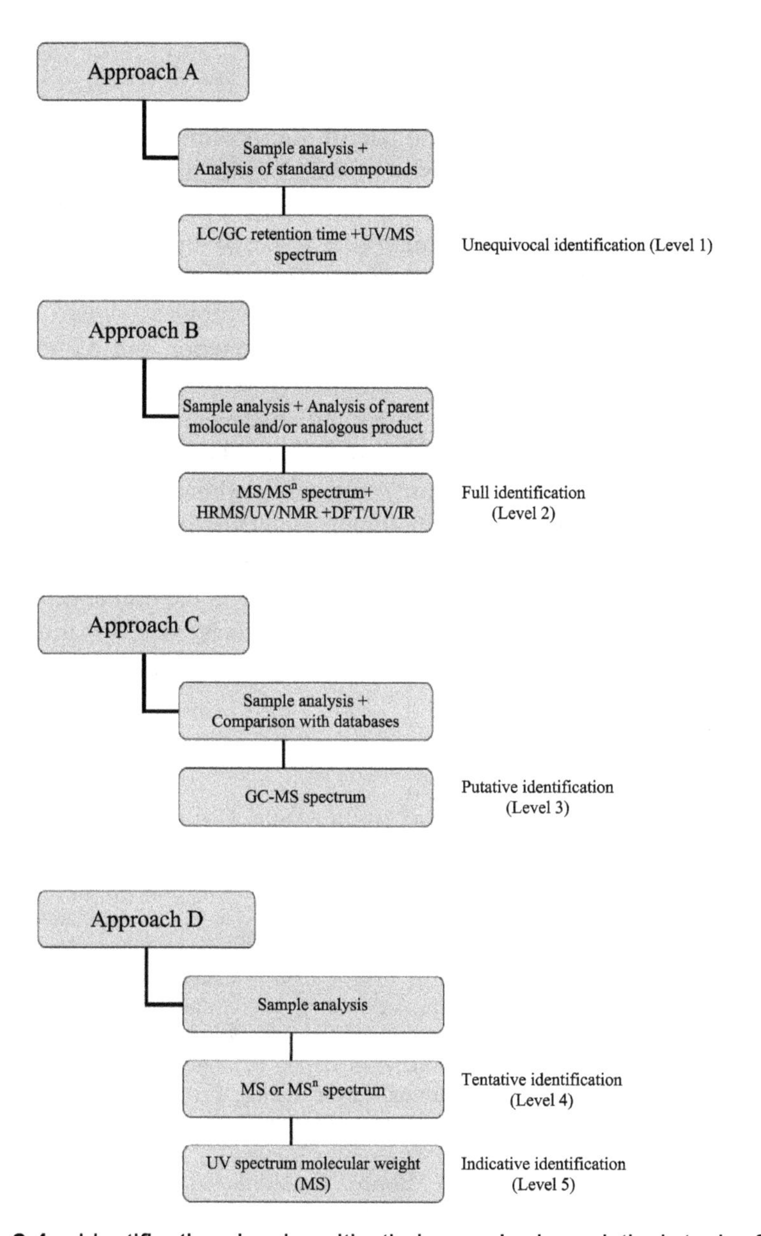

Figure 2.4 Identification levels with their required analytical tools for the identification of transformation products. Adapted form De Witte *et al.* (2009a)

2.5 CONCLUSIONS

The elucidation of the environmental fate of micropollutants is essential due to the potential formation of the persistent metabolites that may cause adverse effect on the environment. The combination of chemical and biological analyses is necessary to evaluate the potential effects posed by the degradation and intermediate products of micropollutants. The developments in analytical instrumental techniques such as the increase in chromatographic resolution, detection sensitivity and selectivity enables the extraction and the detection of target compounds even at nano g concentrations levels from extremely complex matrices. In addition, it is vitally important to develop appropriate tools to identify and quantify unknown degradation products. More sophisticated reactivity-structure linkages that use molecular orbital energy modeling to predict tendencies for neutral micropollutant molecule oxidation by the oxidant (e.g., hydroxyl radical) are currently under development. Generally, the utilization of the standard compounds and databases is recommended, however, the lack of studies for quite many of the micropollutants transformation products triggers the development of the identification methodologies.

2.6 REFERENCES

Armenta, S., Garrigues, S. and de la Guardia, M. (2008) Green Analytical Chemistry. *TrAC Trend. Anal. Chem.* **27**, 497–511.

Bakker, E. and Qin, Y. (2006) Electrochemical Sensors. *Anal. Chem.* **78**, 3965–3984.

Bekbolet, M., Çınar, Z., Kılıç, M., Uyguner, C. S., Minero, C. and Pelizzetti, E. (2009) Photocatalytic oxidation of dinitronaphthalenes: Theory and experiment. *Chemosphere* **75**, 1008–1014.

Cardoza, L. A., Almeida, V. K., Carr, A., Larive, C. K. and Graham, D. W. (2003) Separations coupled with NMR detection. *TrAC Trend. Anal. Chem.* **22**, 766–775.

Carralero, V., Gonzalez-Cortez, A., Yanez-Sedeno, P. and Pingarron, J. M. (2007) Nanostructured progesterone immunosensor using a tyrosinase–colloidal gold–graphite–Teflon biosensor as amperometric transducer. *Anal. Chim. Acta* **596**, 86–91.

Carrier, M., Perol, N., Herrmann, J.-M., Bordes, C., Horikoshi, S., Paisse, J. O., Baudot, R. and Guillard, C. (2006) Kinetics and reactional pathway of Imazapyr photocatalytic degradation Influence of pH and metallic ions. *Appl. Catal. B-Environ.* **65**, 11–20.

Chang, H.-S., Choo, K.-H., Lee, B. and Choi, S.-J. (2009) The methods of identification, analysis, and removal of endocrine disrupting compounds (EDCs) in water. *J. Hazard. Mater.* **172**, 1–12.

Chu, W.-H., Gao, N.-Y., Deng, Y. and Dong, B.-Z. (2009) Formation of chloroform during chlorination of alanine in drinking water. *Chemosphere* **77**, 1346–1351.

De Witte, B., Langenhove, H. V., Demeestere, K. and Dewulf, J. (2009a) Advanced oxidation of pharmaceuticals: Chemical analysis and biological assessment of degradation products. *Crit. Rev. Environ. Sci. Technol.* In Press, Accepted Manuscript.

De Witte, B., Langenhove, H. V., Hemelsoet, K., Demeestere, K., Wispelaere, P. D., Van Speybroeck, V. and Dewulf, J. (2009b) Levofloxacin ozonation in water: Rate determining process parameters and reaction pathway elucidation. *Chemosphere* **76**, 683–689.

Drlica, K. (2001) Antibiotic resistance: Can we beat the bugs? *Drug Discov. Today* **6**, 714–715.

Escandar, G. M., Faber, N. K. M., Goicoechea, H. C., de la Peña, A. M., Olivieri, A. C. and Poppi, R. J. (2007) Second- and third-order multivariate calibration: Data, algorithms and applications. *TrAC Trend. Anal. Chem.* **27**, 752–765.

Exarchou, V., Godejohann, M., Beek, T. A. V., Gerothanassis, I. P. and Vervoort†, J. (2003) LC-UV-Solid-Phase Extraction-NMR-MS combined with a cryogenic flow probe and its application to the identification of compounds present in Greek Oregano. *Anal. Chem.* **75**, 6288–6294.

Exarchou, V., Krucker, M., Beek, T. A. V., Vervoort, J., Gerothanassis, I. P. and Albert, K. (2005) LC-NMR coupling technology: Recent advancements and applications in natural product analysis. *Magn. Reson. Chem.* **43**, 681–687.

Fahnrich, K. A., Pravda, M. and Guilbault, G. G. (2003) Disposable amperometric immunosensor for the detection of polycyclic aromatic hydrocarbons (PAHs) using screen-printed electrodes. *Biosens. Bioelectron.* **18**, 73–82.

Farre, M., Martinez, E. and Barcelo, D. (2007) Validation of interlaboratory studies on toxicity in water samples. *TrAC Trend. Anal. Chem.* **26**, 283–292.

Farre, M., Kantiani, L., Perez, S. and Barcelo, D. (2009) Sensors and biosensors in support of EU Directives. *TrAC Trend. Anal. Chem.* **28**, 170–185.

Farre, M. l., Perez, S., Kantiani, L. and Barcelo, D. (2008) Fate and toxicity of emerging pollutants, their metabolites and transformation products in the aquatic environment. *TrAC Trend. Anal. Chem.* **27**, 991–1007.

Fatta, D., Achilleos, A., Nikolaou, A. and Meric, S. (2007) Analytical methods for tracing pharmaceutical residues in water and wastewater. *TrAC Trend. Anal. Chem.* **26**, 515–533.

Fifield, F. W. and Kealey, D. (2000) *Principles and Practice of Analytical Chemistry*. Cambridge, Blackwell Science Ltd.

Gabet, V., Miege, C., Bados, P. and Coquery, M. (2007) Analysis of estrogens in environmental matrices. *TrAC Trend. Anal. Chem.* **26**, 1113–1131.

Galera, M. M., García, M. D. G. and Goicoechea, H. C. (2007) The application to wastewaters of chemometric approaches to handling problems of highly complex matrices. *TrAC Trend. Anal. Chem.* **26**, 1032–1042.

Garcia-Galan, M. J., Silvia Diaz-Cruz, M. and Barcelo, D. (2008) Identification and determination of metabolites and degradation products of sulfonamide antibiotics. *TrAC Trend. Anal. Chem.* **27**, 1008–1022.

Garrido, M., Rius, F. and Larrechi, M. (2008) Multivariate curve resolution–alternating least squares (MCR-ALS) applied to spectroscopic data from monitoring chemical reactions processes. *Anal. Bioanal. Chem.* **390**, 2059–2066.

Gascon, J., Oubica, A. and Barcely, D. (1997) Detection of endocrine-disrupting pesticides by enzyme-linked immunosorbent assay (ELISA): Application to atrazine. *TrAC Trend. Anal. Chem.* **16**, 554–562.

Gatidou, G., Thomaidis, N. S., Stasinakis, A. S. and Lekkas, T. D. (2007) Simultaneous determination of the endocrine disrupting compounds nonylphenol, nonylphenol ethoxylates, triclosan and bisphenol A in wastewater and sewage sludge by gas chromatography-mass spectrometry. *J. Chromatogr. A* **1138**, 32–41.

Giger, W. (2009) Hydrophilic and amphiphilic water pollutants: Using advanced analytical methods for classic and emerging contaminants. *Anal. Bioanal. Chem.* **393**, 37–44.

Gros, M., Petrovic, M. and Barcelo, D. (2008) Tracing Pharmaceutical Residues of Different Therapeutic Classes in Environmental Waters by Using Liquid Chromatography/Quadrupole-Linear Ion Trap Mass Spectrometry and Automated Library Searching. *Anal. Chem.* **81**, 898–912.

Gu, M. B., Min, J. and Kim, E. J. (2002) Toxicity monitoring and classification of endocrine disrupting chemicals (EDCs) using recombinant bioluminescent bacteria. *Chemosphere* **46**, 289–294.

Hamblen, E. L., Cronin, M. T. D. and Schultz, T. W. (2003) Estrogenicity and acute toxicity of selected anilines using a recombinant yeast assay. *Chemosphere* **52**, 1173–1181.

Hao, C., Zhao, X. and Yang, P. (2007) GC-MS and HPLC-MS analysis of bioactive pharmaceuticals and personal-care products in environmental matrices. *TrAC Trend. Anal. Chem.* **26**, 569–580.

Henry, M. C. and Yonker, C. R. (2006) Supercritical Fluid Chromatography, Pressurized Liquid Extraction, and Supercritical Fluid Extraction. *Anal. Chem.* **78**, 3909–3916.

Hernandez, F., Sancho, J. V., Ibanez, M. and Grimalt, S. (2008) Investigation of pesticide metabolites in food and water by LC-TOF-MS. *TrAC Trend. Anal. Chem.* **27**, 862–872.

Hernandez, F., Sancho, J. V., Ibanez, M. and Guerrero, C. (2007) Antibiotic residue determination in environmental waters by LC-MS. *TrAC Trend. Anal. Chem.* **26**, 466–485.

Hernando, M. D., Gómez, M. J., Agüera, A. and Fernández-Alba, A. R. (2007) LC-MS analysis of basic pharmaceuticals (beta-blockers and anti-ulcer agents) in wastewater and surface water. *TrAC Trend. Anal. Chem.* **26**, 581–594.

Hogenboom, A. C., van Leerdam, J. A. and de Voogt, P. (2009) Accurate mass screening and identification of emerging contaminants in environmental samples by liquid chromatography-hybrid linear ion trap Orbitrap mass spectrometry. *J. Chromatogr. A* **1216**, 510–519.

Hyotylainen, T. and Hartonen, K. (2002) Determination of brominated flame retardants in environmental samples. *TrAC Trend. Anal. Chem.* **21**, 13–30.

Ibanez, M., Sancho, J. V., Hernandez, F., McMillan, D. and Rao, R. (2008) Rapid non-target screening of organic pollutants in water by ultraperformance liquid chromatography coupled to time-of-light mass spectrometry. *TrAC Trend. Anal. Chem.* **27**, 481–489.

Katsumata, H., Kaneco, S., Suzuki, T., Ohta, K. and Yobiko, Y. (2006) Degradation of polychlorinated dibenzo-p-dioxins in aqueous solution by Fe(II)/H2O2/UV system. *Chemosphere* **63**, 592–599.

Kormos, J. L., Schulz, M., Wagner, M. and Ternes, T. A. (2009) Multistep Approach for the Structural Identification of Biotransformation Products of Iodinated X-ray Contrast Media by Liquid Chromatography/Hybrid Triple Quadrupole Linear Ion Trap Mass Spectrometry and 1H and 13C Nuclear Magnetic Resonance. *Anal. Chem.* **81**, 9216–9224.

Korner, W., Bolz, U., Sussmuth, W., Hiller, G., Schuller, W., Hanf, V. and Hagenmaier, H. (2000) Input/output balance of estrogenic active compounds in a major municipal sewage plant in Germany. *Chemosphere* **40**, 1131–1142.

Kosjek, T. and Heath, E. (2008) Applications of mass spectrometry to identifying pharmaceutical transformation products in water treatment. *TrAC Trend. Anal. Chem.* **27**, 807–820.

Kostopoulou, M. and Nikolaou, A. (2008) Analytical problems and the need for sample preparation in the determination of pharmaceuticals and their metabolites in aqueous environmental matrices. *TrAC Trend. Anal. Chem.* **27**, 1023–1035.

Kot-Wasik, A., Debska, J. and Namiesnik, J. (2007) Analytical techniques in studies of the environmental fate of pharmaceuticals and personal-care products. *TrAC Trend. Anal. Chem.* **26**, 557–568.

Kralj, M. B., Trebse, P. and Franko, M. (2007) Applications of bioanalytical techniques in evaluating advanced oxidation processes in pesticide degradation. *TrAC Trend. Anal. Chem.* **26**, 1020–1031.

Kucharska, M. and Grabka, J. (2010) A review of chromatographic methods for determination of synthetic food dyes. *Talanta* **80**, 1045–1051.

Lange, F., Cornelissen, S., Kubac, D., Sein, M. M., von Sonntag, J., Hannich, C. B., Golloch, A., Heipieper, H. J., Möder, M. and von Sonntag, C. (2006) Degradation of macrolide antibiotics by ozone: A mechanistic case study with clarithromycin. *Chemosphere* **65**, 17–23.

Lee, B.-D., Iso, M. and Hosomi, M. (2001) Prediction of Fenton oxidation positions in polycyclic aromatic hydrocarbons by Frontier electron density. *Chemosphere* **42**, 431–435.

Lidman, U. (2005). The nature and chemistry of toxicants. Environmental Toxicity Testing. Thompson, K. C., Wadhia, K. and Loibner, A. P. Oxford, UK, Blackwell Publishing.

Liu, G., Li, X., Zhao, J., Horikoshi, S. and Hidaka, H. (2000) Photooxidation mechanism of dye alizarin red in TiO2 dispersions under visible illumination: An experimental and theoretical examination. *J. Mol. Catal. A-Chem.* **153**, 221–229.

Liu, S., Yuan, L., Yue, X., Zheng, Z. and Tang, Z. (2008) Recent Advances in Nanosensors for Organophosphate Pesticide Detection. *Adv. Powder Tech.* **19**, 419–441.

Lopez de Alda, M. J., Díaz-Cruz, S., Petrovic, M. and Barceló, D. (2003) Liquid chromatography-(tandem) mass spectrometry of selected emerging pollutants (steroid sex hormones, drugs and alkylphenolic surfactants) in the aquatic environment. *J. Chromatogr. A* **1000**, 503–526.

Maciá, A., Borrull, F., Calull, M. and Aguilar, C. (2007) Capillary electrophoresis for the analysis of non-steroidal anti-inflammatory drugs. *TrAC Trend. Anal. Chem.* **26**, 133–153.

Madej, K. (2009) Microwave-assisted and cloud-point extraction in determination of drugs and other bioactive compounds. *TrAC Trend. Anal. Chem.* **28**, 436–446.

McDowell, D. C., Huber, M. M., Wagner, M., Gunten, U. V. and Ternes, T. A. (2005) Ozonation of carbamazepine in drinking water: Identification and kinetic study of major oxidation products. *Environ. Sci. Technol.* **39**, 8014–8022.

Miege, C., Bados, P., Brosse, C. and Coquery, M. (2009) Method validation for the analysis of estrogens (including conjugated compounds) in aqueous matrices. *TrAC Trend. Anal. Chem.* **28**, 237–244.

Moco, S., Vervoort, J., Moco, S., Bino, R. J., De Vos, R. C. H. and Bino, R. (2007) Metabolomics technologies and metabolite identification. *TrAC Trend. Anal. Chem.* **26**, 855–866.

Murk, A. J., Legler, J., Lipzig, M. M. v., Meerman, J. H., Belfroid, A. C., Spenkelink, A., Burg, B. V. D. and G.B. Rijs, D. V. (2002) Detection of estrogenic potency in wastewater and surface water with three in vitro bioassays. *Environ. Toxicol. Chem.* **21**, 16–21.

Nelson, J., Bishay, F., Roodselaar, A. v., Ikonomou, M. and Law, F. C. P. (2007) The use of in vitro bioassays to quantify endocrine disrupting chemicals in municipal wastewater treatment plant effluents. *Sci. Total Environ.* **374**, 80–90.

Notsu, H., Tatsuma, T. and Fujishima, A. (2002) Tyrosinase-modified boron-doped diamond electrodes for the determination of phenol derivatives. *J. Electroanal. Chem.* **523**, 86–92.

Ohko, Y., Iuchi, K.-i., Niwa, C., Tatsuma, T., Nakashima, T., Iguchi, T., Kubota, Y. and Fujishima, A. (2002) 17β-Estradiol degradation by TiO2 photocatalysis as a means of reducing estrogenic activity. *Environ. Sci. Technol.* **36**, 4175–4181.

Oliveira, D. P., Carneiro, P. A., Sakagami, M. K., Zanoni, M. V. B. and Umbuzeiro, G. A. (2007) Chemical characterization of a dye processing plant effluent−Identification of the mutagenic components. *Mutat. Res.-Gen.Tox. En.* **626**, 135–142.

Parr, R. G. and Weitao, Y. (1994) *Density-Functional Theory of Atoms and Molecules.* Oxford, USA, Oxford Univercity Press.

Pavlovic, D. M., Babic, S., Horvat, A. J. M. and Kastelan-Macan, M. (2007) Sample preparation in analysis of pharmaceuticals. *TrAC Trend. Anal. Chem.* **26**, 1062–1075.

Petroviciu, I., Albu, F. and Medvedovici, A. (2010) LC/MS and LC/MS/MS based protocol for identification of dyes in historic textiles. *Microchemical. J.* In Press, Accepted Manuscript.

Pico, Y., Rodriguez, R. and Manes, J. (2003) Capillary electrophoresis for the determination of pesticide residues. *TrAC Trend. Anal. Chem.* **22**, 133–151.

Pillon, A., Boussioux, A. M., Escande, A., Ait-Aissa, S., E. Gomez, Fenet, H., Ruff, M., Moras, D., Vignon, F., Duchesne, M. J., Casellas, C., Nicolas, J. C. and Balaguer, P. (2005) Binding of estrogenic compounds to recombinant estrogen receptor-alpha: Application to environmental analysis. *Environ. Health Perspect.* **113**, 278–284.

Pinheiro, H. M., Touraud, E. and Thomas, O. (2004) Aromatic amines from azo dye reduction: Status review with emphasis on direct UV spectrophotometric detection in textile industry wastewaters. *Dyes Pigments* **61**, 121–139.

Poiger, T., Richardson, S. D. and Baughman, G. L. (2000) Identification of reactive dyes in spent dyebaths and wastewater by capillary electrophoresis-mass spectrometry. *J. Chromatogr. A* **886**, 271–282.

Radjenovic, J., Petrovic, M. and Barceló, D. (2009) Complementary mass spectrometry and bioassays for evaluating pharmaceutical-transformation products in treatment of drinking water and wastewater. *TrAC Trend. Anal. Chem.* **28**, 562–580.

Radjenovic, J., Petrovic, M., Barceló, D. and Petrovic, M. (2007) Advanced mass spectrometric methods applied to the study of fate and removal of pharmaceuticals in wastewater treatment. *TrAC Trend. Anal. Chem.* **26**, 1132–1144.

Raman Suri, C., Boro, R., Nangia, Y., Gandhi, S., Sharma, P., Wangoo, N., Rajesh, K. and Shekhawat, G. S. (2009) Immunoanalytical techniques for analyzing pesticides in the environment. *TrAC Trend. Anal. Chem.* **28**, 29–39.

Raynie, D. E. (2004) Modern Extraction Techniques. *Anal. Chem.* **76**, 4659–4664.

Reemtsma, T., Quintana, J. B., Rodil, R., GarcIía-López, M. and Rodríguez, I. (2008) Organophosphorus flame retardants and plasticizers in water and air I. Occurrence and fate. *TrAC Trend. Anal. Chem.* **27**, 727–737.

Richardson, S. D. (2000) Environmental Mass Spectrometry. *Anal. Chem.* **72**, 4477–4496.

Richardson, S. D. (2003) Disinfection by-products and other emerging contaminants in drinking water. *TrAC Trend. Anal. Chem.* **22**, 666–684.
Richardson, S. D. (2008) Environmental Mass Spectrometry: Emerging Contaminants and Current Issues. *Anal. Chem.* **80**, 4373–4402.
Richardson, S. D. (2009) Water Analysis: Emerging Contaminants and Current Issues. *Anal. Chem.* **81**, 4645–4677.
Richardson, S. D., Plewa, M. J., Wagner, E. D., Schoeny, R. and DeMarini, D. M. (2007) Occurrence, genotoxicity, and carcinogenicity of regulated and emerging disinfection by-products in drinking water: A review and roadmap for research. *Mutat. Res.-Rev. Mutat.* **636**, 178–242.
Roh, H., Subramanya, N., Zhao, F., Yu, C.-P., Sandt, J. and Chu, K.-H. (2009) Biodegradation potential of wastewater micropollutants by ammonia-oxidizing bacteria. *Chemosphere* **77**, 1084–1089.
Routledge, E. J. and Sumpter, J. P. (1997) Structural features of alkyl-phenolic chemicals associated with estrogenic activity. *J. Biol. Chem.* **272**, 3280–3288.
Rubio, S. and Perez-Bendito, D. (2009) Recent Advances in Environmental Analysis. *Anal. Chem.* **81**, 4601–4622.
Salste, L., Leskinen, P., Virta, M. and Kronberg, L. (2007) Determination of estrogens and estrogenic activity in wastewater effluent by chemical analysis and the bioluminescent yeast assay. *Sci. Total Environ.* **3**, 343–351.
Skladal, P. (1999) Effect of methanol on the interaction of monoclonal antibody with free and immobilized atrazine studied using the resonant mirror-based biosensor. *Biosens. Bioelectron.*, 257–263.
Streck, G. (2009) Chemical and biological analysis of estrogenic, progestagenic and androgenic steroids in the environment. *TrAC Trend. Anal. Chem.* **28**, 635–652.
Sumner, L., Amberg, A., Barrett, D., Beale, M., Beger, R., Daykin, C., Fan, T., Fiehn, O., Goodacre, R., Griffin, J., Hankemeier, T., Hardy, N., Harnly, J., Higashi, R., Kopka, J., Lane, A., Lindon, J., Marriott, P., Nicholls, A., Reily, M., Thaden, J. and Viant, M. (2007) Proposed minimum reporting standards for chemical analysis. *Metabolomics* **3**, 211–221.
Taranova, L. A., Fesay, A. P., Ivashchenko, G. V., Reshetilov, A. N., Winther-Nielsen, M. and Emneus, J. (2004) Comamonas testosteroni Strain TI as a Potential Base for a Microbial Sensor Detecting Surfactants *Appl. Biochem. Microbiol.* **40**, 404–408.
Tobiszewski, M., Mechlinska, A., Zygmunt, B. and Namiesnik, J. (2009) Green analytical chemistry in sample preparation for determination of trace organic pollutants. *TrAC Trend. Anal. Chem.* **28**, 943–951.
Tschmelak, J., Kumpf, M., Kappel, N., Proll, G. and Gauglitz, G. (2006) Total internal reflectance fluorescence (TIRF) biosensor for environmental monitoring of testosterone with commercially available immunochemistry: Antibody characterization, assay development and real sample measurements. *Talanta* **69**, 343.
Veith, G. D., Greenwood, B., Hunter, R. S., Niemi, G. J. and Regal, R. R. (1988) On the intrinsic dimensionality of chemical structure space. *Chemosphere* **17**, 1617–1630.
Vogna, D., Marotta, R., Napolitano, A. and d'Ischia, M. (2002) Advanced oxidation chemistry of paracetamol. UV/H2O2-induced hydroxylation/degradation pathways and 15N-aided inventory of nitrogenous breakdown products. *J. Org. Chem.* **67**, 6143–6151.
Wadhia, K. and Thompson, K. C. (2007) Low-cost ecotoxicity testing of environmental samples using microbiotests for potential implementation of the Water Framework Directive. *TrAC Trend. Anal. Chem.* **26**, 300–307.

Wishart, D. S. (2008) Quantitative metabolomics using NMR. *TrAC Trend. Anal. Chem.* **27**, 228–237.

Wolska, L., Sagajdakow, A., Kuczynska, A. and Namiesnik, J. (2007) Application of ecotoxicological studies in integrated environmental monitoring: Possibilities and problems. *TrAC Trend. Anal. Chem.* **26**, 332–344.

Young, D. C. (2001) *COMPUTATIONAL CHEMISTRY:A Practical Guide for Applying Techniques to Real-World Problems*, John Wiley & Sons, Inc.

Zhu, X., Feng, X., Yuan, C., Cao, X. and Li, J. (2004) Photocatalytic degradation of pesticide pyridaben in suspension of TiO2: Identification of intermediates and degradation pathways. *J. Mol. Catal. A-Chem.* **214**, 293–300.

3

EDC Sensors and Biosensors

Achintya N. Bezbaruah and Harjyoti Kalita

3.1 INTRODUCTION

Endocrine systems control hormones and activity-related hormones in many living organisms including mammals, birds, and fish. The endocrine system consists of various glands located throughout the body, hormones produced by the glands, and receptors in various organs and tissues that recognize and respond to the hormones (USEPA, 2010a). There are some chemicals and compounds that cause interferences in the endocrine system and these substances are known as endocrine disrupting chemicals (EDCs). Wikipedia states that EDCs or "endocrine disruptors are exogenous substances that act like hormones in the endocrine system and disrupt the physiologic function of endogenous hormones. They are sometimes also referred to as hormonally active agents." EDCs can be man-made or natural. These compounds are found in plants (phytochemicals), grains, fruits and vegetables, and fungus. Alkyl-phenols found in detergents, bisphenol A used in PVC products, dioxins, various drugs, synthetic estrogens found in birth control pills, heavy metals (Pb, Hg, Cd), pesticides, pasticizers, and phenolic products are all examples of EDCs from a long list that is rapidly getting longer. It is suspected that EDCs could be harmful to living organisms, therefore, there is a concerted effort to detect and treat EDCs before they can cause harm to the ecosystem components.

In this chapter we are discussing some of the EDC sensors and biosensors which have been developed over the last few years. The first part of the chapter is dedicated to EDC sensors and biosensors. We then include other sensors while discussing trends in sensors and biosensors keeping in mind that the technology used for the other sensors can be very well adapted for the fabrication of EDC sensors. The purpose of this chapter is to offer an opportunity to the readers to have a feel of the enormous possibilities that sensor and biosensor technologies hold for detecting and quantifying micro-pollutants in the environment. The chapter is based on a number of original and review papers which are cited throughout the chapter.

3.2 SENSORS AND BIOSENSORS

3.2.1 The need for alternative methods

The most widely used methods for the determination of various EDCs are high-performance liquid chromatography (HPLC), liquid chromatography coupled with electrochemical detection (LC-ED), liquid chromatography coupled with mass spectrometry (LC-MS), capillary electrophoresis (CE), gas chromatography (GC), and gas chromatography coupled with mass spectrometry (GC–MS) (Nakata *et al.,* 2005; Petrovic *et al.,* 2005; Liu *et al.,* 2006a; Vieno *et al.,* 2006; Wen *et al.,* 2006; Gatidou *et al.,* 2007; Comerton *et al.,* 2009; Mottaleb *et al.,* 2009). These methods offer excellent selectivity and detection limits, however, they are not suitable for rapid processing of multiple samples and real-time detection. They involve highly trained operators, time-consuming detection processes, and complex pre-treatment steps. The instruments are sophisticated and expensive. Further, the methods are unsuitable for field studies and in-situ monitoring of samples (Rahman *et al.,* 2007; Rodrigues *et al.,* 2007; Huertas-Perez and Garcia-Campana, 2008; Saraji and Esteki, 2008; Blăzkova *et al.,* 2009; Le Blanc *et al.,* 2009; Suri *et al.,* 2009; Yin *et al.,* 2009). EDCs can also be detected using immunochemical techniques like enzyme-linked immunosorbent assays (ELISA) (Marchesini *et al.*, 2005, 2007; Rodriguez-Monaz *et al.*, 2005; Kim *et al.*, 2007), however, these immunotechniques are less advantageous than chromatographic techniques because the stability of the biological materials used in assays is lower, and the assays involve complicated multistage steps that may involve expensive equipment. The specific antibodies or particular proteins must be obtained by recombinant techniques for assay fabrication (Le Blanc *et al.*, 2009; Yin *et al.*, 2009). ELISA-based methods are difficult to use by non-specialized laboratories and in the field. They involve labor intensive operations like repeated incubation and washing, and enzyme reaction for final signal generation (Blăzkova *et al.*, 2009). Further, ELISA-based methods are specific for a single compound or, at the best, its structurally related compounds. They can not be used for multi-analyte detection and quantification. Unfortunately EDCs are structurally diverse, and there is a continuous introduction of new EDCs into the environment due to market driven evolution of chemicals (Marchesini *et al.*, 2007).

There is a need for new, simple analytical techniques for EDCs with reliable and fast responses. High cost is a major hindrance for the introduction of new tools and equipment into existing laboratories. Lower capital, operation, and maintenance costs will make such equipment very attractive. The equipment should be simple to operate, less time consuming, have high sensitivity, and capable of real-time detection. Sensors of various kinds can be the alternative for expensive analytical methods (Yin *et al.*, 2009). The high number and structural diversity of EDCs calls for the urgent development of sensors for monitoring activities (or measuring effects of the EDCs) rather than only the concentration of a single, or a set of, compounds (Le Blanc *et al.*, 2009).

3.2.2 Electrochemical sensors

Electrochemical sensors are cheap, simple to fabricate, and reusable. They have high stability and sensitivity. They can potentially be used for other species with the necessary modifications (Kamyabi and Aghajanloo, 2008; Yin *et al.*, 2009). Many phenolic compounds are successfully detected using electrochemical sensors as most sensors are oxidized at readily accessible potentials (Liu *et al.*, 2005a). Being able to decrease the redox potential needed for the electro-chemical reaction makes the sensor more adaptable and sensitive to other EDCs. Chemically modified carbon paste electrodes have been prepared by Yin *et al.* (2009) for the detection of bisphenol A (BPA). Cobalt phthalocyanine modifier has been used in electrodes to help decrease the redox potential. Increased sensitivity and selectivity have been achieved for BPA in an aqueous medium. The detection limit was 1.0×10^{-8} M (Yin *et al.*, 2009).

3.2.3 Biosensors

While chemicals and electrochemical strategies for determining contaminants are robust, they don't give us a complete picture of the ecological risks involved and impacts observed. Such information can be obtained only after proper interpretation by experts. However, combining both biological responses and chemical analyses may give us a better picture of the situation. We should be able to get results for the identification of toxic hotspots, toxic chemical characterization, and estimation of ecological risks of the contaminants at relevant spatial scales. Such assessments call for rapid, inexpensive screening to characterize the extent of the contamination (Brack *et al.,* 2007; Farré *et al.,* 2007; Blasco and Picó, 2009; Fernandez *et al.,* 2009; USEPA, 2010b). Further, the use of biological tools will help in the quantification of an EDC or any other pollutant in terms of its eco-effects (Marchesini *et al.,* 2007). Different biological tools including biosensors have been extensively used in recent years. Biomonitoring is becoming an essential component in effective environmental monitoring (Grote *et al.,* 2005; Rodriguez-Mozaz *et al.,* 2005; Barcelo and Petrovic, 2006; Gonzalez-Doncel *et al.,* 2006; González-Martinez *et al.,* 2007; Tudorache and Bala, 2007; Blasco and Picó, 2009).

A biosensor is defined by the International Union of Pure and Applied Chemistry (IUPAC) as "a device that uses specific biochemical reactions mediated by isolated enzymes, immunosystems, tissues, organelles or whole cells to detect chemical compounds usually by electrical, thermal or optical signals." Organelles include both mitochondria and chloroplasts (where photosynthesis takes place). Biosensors offer a number of advantages over conventional analytical techniques including portability, miniaturization, and on-site monitoring. They are also capable of measuring pollutants in complex matrices and with minimal sample preparation. Even though biosensors can't yet measure analytes as accurately as conventional analytical methods they are very good tools for routine testing and screening (Rodriguez-Mozaz *et al.,* 2006a). Rodriguez-Mozaz *et al.* (2006a) have carried out an extensive review of biosensors for environmental analysis and monitoring. Their review covers biosensors for the measurement of pesticides, hormones, PCBs, dioxins, bisphenol A, antibiotics, phenols, and EDC effects. The monitoring process using conventional analytical methods involves the collection of water samples followed by laboratory-based instrumental analysis, and such analyses only provide snapshots of the situation at the sampling site and time rather than more realistic information on spatio-temporal variations in water characteristics (Allan *et al.,* 2006; Rodriguez-Mozaz *et al.,* 2006a). Biosensors can be useful in situations when continuous and spatial data are needed. Biosensors have high specificity and sensitivity. Further, a biosensor can not only determine chemicals of concern but can record their biological effects (toxicity, cytotoxicity, genotoxicity or endocrinedisrupting effects). Often, information on biological effects is more relevant than the chemical composition. A biosensor can provide an assessment of both the total and bioavailable/bioaccesible contaminants. However, the majority of the biosensor systems developed is still in the lab tables or in prototype stages and needs to be validated before mass production and use (Rodriguez-Mozaz *et al.,* 2006a; Farré *et al.,* 2009a)

Rodriguez-Mozaz *et al.* (2006a) further discussed specific biosensors developed for certain contaminants. Organophosphorous hydrolase (OPH) can be combined with optical or amperometric transducers to measure absorbance or oxidation reduction currents generated by hydrolysis byproduct of many pesticides (e.g., paraoxon, parathion) and chemical warfare agents (e.g., sarin and soman). OPH [or Phosphotriesterase (PTE)] enzyme can hydrolyze organophosphate pesticides to release p-nitrophenol which is electroactive and chromophoric and can, thus, be measured with an OPH biosensor (Rodriguez-Mozaz *et al.,* 2006a).

EDCs bind to a hormone receptor site or a transport protein and express their biological effects. This can interpreted as (1) mimicking or antagonizing the effects of the endogenous hormone; or (2) disrupting the synthesis or metabolism of endogenous hormones or hormone receptors. It is possible to use the same receptors or transport proteins targeted by the EDCs as their bio-recognition elements. A method like this allows us to monitor the endocrine disrupting potency of single or multiple chemicals in a sample based on their bio-effect(s) on the receptor (Marchesini *et al.,* 2007). For example, human estrogen receptor α (ERα) group is capable of interacting with a large variety of chemicals (e.g., phytoestrogens, xenoestrogens, pesticides) that cause estrogenic effects *in-vivo.* The receptor family offers a variety of opportunities for use in tailor-made applications for EDCs. These include interaction between the ligand-binding domain (LBD) and its ligands or peptides derived from co-activator or co-repressor proteins (Fechner *et al.,* 2009), and the interaction between DNA-binding domains and certain DNA sequences (estrogen response elements) (Asano *et al.,* 2004; Le Blanc *et al.,* 2009). EDCs are chemicals that are able to interfere with interactions between ERα and these domains. Le Blanc *et al.* (2009) used this knowledge of the effects of EDCs on ERα receptors. They labeled ERα in an assay to determine the impact of EDCs. As compared to conventional methods, the new assays can determine the total effect on the receptor instead of concentrations of single compounds. The signal obtained is the response of the organism which is exposed to EDCs. The detection limit was reported to be 0.139 nM of estradiol equivalents. While standard analytical techniques are designed to find only known compounds, the results of this assay incorporate all known and unknown EDCs (and possibly other compounds). These data are difficult to compare and validate. While validation will be a necessary step in the near future, the assay can be used now to monitor changes in the estrogenicity of environmental samples over time. Sanchez-Acevedo *et al.* (2009) recently reported the detection of picomolar concentrations of bisphenol A (BPA) in water using a carbon nanotube field-effect transistor (CNTFET). The CNTFET is functionalized with ERα where ERα serves as the recognition layer for the sensor. The sensor uses the molecular recognition principles. Single-walled carbon nanotubes (SWCNTs) have been used as transducers and ERα is adsorbed onto their surface. A blocking agent has been used in order to avoid non-specific adsorption on the SWCNT surface. BPA has been detected up to 2.19×10^{-12} M in aqueous solution in 2 minutes. Fluoranthene, pentacloronitrobenzene, and malathion present in the water didn't produce any interferences. Such a biosensor can be useful in a label-free platform for detecting other analytes by using an appropriate nuclear receptor (Sanchez-Acevedo *et al.*, 2009).

Marchesini *et al.* (2006) reported the use of a plasmon resonance (SPR)-based label-free biosensor manufactured by an US manufacturer. They used this in combination with a ready-to-use biosensor chip to screen bio-effect related molecules and predicted possible SPR biosensor uses for EDC bio-effect monitoring. While it is possible to use such biosensors for EDC detection, the exorbitant price of commercially available systems and the lack of portability for in-situ analysis are the major drawbacks of SPR-based biosensors. These are the major challenges that need to be overcome for SPR-based biosensors to be popular (Marchesini *et al.*, 2007). SPR-based sensors have been used for dioxins, polychlorinated biphenyl and atrazine (Farré *et al.*, 2009b) and the sensor needed 15 min for a single sample measurement. A portable SPR immunosensor for organophosphate pesticide chlorpyrifos (detection limit of 45–64 ng/L) as well as single and multi-analyte SPR assays for the simultaneous detection of cholinesterase-inhibiting pesticides have been reported (Mauriz *et al.*, 2006a,b; Farré *et al.*, 2009b). These sensors were made re-usable through the formation of alkanethiol self-assembled monolayers.

Bacterial and other cells are also used in sensors known as whole-cell sensors. During ongoing research to detect the estrogenic properties of commonly used chemicals, products, and their ingredients, researchers have developed many different live animal, whole cell, and *in-vitro* binding assays. ER-positive breast cancer cell lines show increased proliferation due to estrogenic activity. Many *in-vitro* assays can be used to detect estrogens. Hormone responsive reporter assays in human breast cancer cells and rat fibroblast cells are examples of this. However, they typically need complex equipment and reagents, and are highly sensitive to interferences (Gawrys *et al.,* 2009). Gawrys *et al.* (2009) have developed a simple detection system in which the ligand-binding domain of the estrogen receptor β (ERβ) has been incorporated into a larger allosteric reporter protein in *E. coli* cells. The reporter protein expresses itself by creating a hormone-dependent growth phenotype in thymidylate synthase deficient *E. coli* strains. If a knockout media is used then there will be a marked change in growth in the presence of various test compounds that can be detected by a simple measure of the turbidity. Estrogenic behavior in compounds found in consumer products was tested using this technique. The allosteric biosensor *E. coli* strain was used to evaluate estrogenicity of a variety of compounds and complex mixtures used in common consumer products. Perfumes, hand and body washes, deodorants, essential oils and herbal supplements were included in the samples tested with 17 β-estradiol and two thyroid hormones (as controls). The system offered an additional advantage of detecting cytotoxicity of various compounds to the sensor strains. Cytotoxicity detection was based on the loss of viability of the cells in the presence of the test compound under nonselective conditions (Skretas and Wood 2005; Gawrys *et al.,* 2009).

A review by Farré *et al.* (2009) covered the developments in whole-cell biosensors. Amperometric biosensors based on genetically-engineered *Moraxella sp.* and *Pseudomonas putida* JS444 with surface-expressed OPH were used for the detection of organophosphorous pesticides (Lei *et al.,* 2005, 2007). The sensors measured up to 277 ng/L of fenitrothion. Liu *et al.* (2007) used horizontally aligned SWCNTs to fabricate biosensors for the real-time detection of organophosphate. SWCNT surface immobilized OPH triggers enzymatic hydrolysis of pesticides (e.g., paraoxon). The hydrolysis causes a detectable change in the conductance of the SWCNTs which is correlated to the organophosphorous pesticide concentration. Glass electrodes have been modified with genetically-engineered *E. coli* and organophosphorous pesticides degrading bacteria *Flavobacteium sp.* (Mulchandani *et al.,* 1998a, b; Berlein *et al.,* 2002).

The photosynthesis reaction mechanism (photosystem II or PS II) has also been used in biosensors (Giardi and Pace, 2005; Campàs *et al.,* 2008). The PSII-based biosensors can recognize analytes such as triazines, phenylurea, diazines, and phenolic compounds. In PSII, light is first absorbed by chlorophyll–protein complexes. The photochemically active reaction centre chlorophyll (P680) then becomes excited and donates electrons to the primary pheophytin acceptor. This charge separation is stabilized by the transfer of an electron to quinone Q_A and subsequently to Q_B. Q_A, a firmly bound plastoquinone molecule is located in the D2 subunit while Q_B is a mobile plastoquinone located in the D1 subunit of PSII. Q_A and Q_B are binding pockets. Many herbicides can bind reversibly to the "herbicide-binding niche" which is the D1 subunit of PSII within its Q_B binding pocket. Once bound to the niche, the herbicides displace the plastoquinone Q_B and inhibit natural electron transfer. Once the electron flow is stopped, oxygen evolution also stops and the fluorescence properties of PSII change (Giardi and Pace, 2006; Chaplen *et al.*, 2007; Campàs *et al.*, 2008). While this is an exciting way of detecting herbicides, PSII based herbicide recognition is not very dependable as heavy metals may interfere (Chaplen *et al.*, 2007). Such interferences limit the use of PSII based biosensors (Giardi *et al.*, 2009). There are about 65 amino acids in a herbicide binding site. Giardi *et al.* (2009) hypothesized that modifying only one amino acid within the Q_B binding pocket would change photosynthetic activity and herbicide-binding characteristics considerably. Also, depending on the position and type of amino acid substitution, different herbicides will show affinity for the site. They used unicellular green algae *Chlamydomonas reinhardtii* strains and modified the Q_B pocket. The mutant algae cells were then used to fabricate a re-usable and portable optical biosensor with enhanced sensitivity toward different herbicide (e.g., atrazine, diuron, linuron). The detection limits ranged from 0.9×10^{-11} to 3.0×10^{-9} M (Giardi *et al.*, 2009).

3.2.4 New generation immunosensors

While conventional ELISA is considered inadequate for many contaminants, new generation of immunosensors are becoming increasingly popular. Electrochemical immunosensors can be used for real-time in-situ monitoring of EDCs like BPA. Immunosensors are widely used for the detection of an analyte where an enzyme is labeled with a specific antigen. Enzyme labeling is a time consuming and complicated procedure. However, label-free electrochemical immunosensors represent a very attractive technique to detect EDCs by monitoring changes in electronic properties due to immunocomplex formation on the electrode surface (Rahman *et al.*, 2007). Rahman *et al.* (2007) have fabricated a label-free impedimetric immunosensor for the direct detection of BPA. They prepared antigens through the conjugation of BHPVA with bovine serum albumin and then produced a specific polyclonal antibody. A covalent immobilization technique was used during sensor fabrication to attach a polyclonal antibody onto a carboxylic acid group which was functionalized on nanoparticle-based conducting polymer (Rahman *et al.*, 2005) coated onto a glassy carbon electrode. Silver-Silver Chloride (Ag/AgCl) and platinum (Pt) wires were used as reference and counter electrodes, respectively. The detection limit was determined to be 0.3 μg BPA/L (Rahman *et al.*, 2007).

Suri *et al.* (2009) provide an elaborate discussion of immunoanalytical techniques for pesticide analysis. Immunochemical techniques offer great potential for developing inexpensive, reliable sensors for effective field monitoring of many toxic molecules. Such a sensor can be based on the specificity of the antibody–antigen (A_b–A_g) reaction. Specific antibodies can be produced against pesticide molecules. As compared to other sensors, immunosensors can provide quantitative results with similar or even greater sensitivity, accuracy, and precision. Immunosensor data are comparable to standard chemical-based methods as well. Immunosensors are important tools because they complement existing analytical methods and provide low-cost confirmatory tests for many compounds, including pharmaceuticals and pesticides. A detectable signal is obtained from binding interactions between immobilized biomolecules (A_b or A_g) and analytes (A_g or A_b) of interest. Immobilization typically happens on the transducer surface. The sensor is based on the molecular recognition characteristics and, hence, the high selectivity of an A_b can be achieved (Farré *et al.*, 2007; Suri *et al.*, 2009). A_b binds reversibly with a specific A_g in a solution to form an immunocomplex ($A_b - A_g$) (Suri *et al.*, 2009):

$$A_b + A_g \frac{K_a}{K_d} \leftrightarrow A_b - A_g \tag{3.1}$$

where K_a = rate constants for association and K_d = rate constants for dissociation. The equilibrium constant (or the affinity constant) of the reaction is:

$$K = \frac{K_a}{K_d} = \frac{[A_b - A_g]}{[A_b][A_g]} \tag{3.2}$$

An immunocomplex typically has a low K_d value (in the range 10^{-6}–10^{-12}) and displays a high K value (~104). The equilibrium kinetics in solution suggest rapid association and dissociation while the direction of equilibrium depends on the overall affinity (Suri *et al.*, 2009). Immunosensors were initially used for clinical diagnostics. The development and applications of immunosensors for environmental pollutants (e.g., pesticides) are relatively new. The reasons for the time lag include difficulty in finding antibodies against pesticides (pesticides being low in molecular weight) (Suri *et al.*, 2009). A_b affinity and specificity primarily determines the analytical capability of an immunosensor and, hence, the development of antibodies represents a key step in the sensor development (Farré *et al.*, 2007). Now, that antibodies can be produced against low-molecular mass pesticides, immunosensors are expected to become cost-effective devices for the on-site monitoring of pesticides. However, many challenges remain. One of these challenges was the development of pesticide species-specific immunosensors. Pesticides are usually nonimmunogenic and it is, therefore, crucial to synthesize a suitable hapten molecule which can be coupled with a carrier protein to make a stable carrier-hapten complex. The carrier-hapten conjugate should mimic the structure of small pesticide molecules such that a suitable A_b for a particular immunoassay for the specific target molecule can be generated (Suri *et al.*, 2009).

With recent developments, immunoassays have been used for the measurement of both single and multiple analytes. Organic pollutants like pesticides, polychlorinated biphenyls (PCBs), and surfactants can be rapidly and efficiently determined through immunochemistry (Farré *et al.*, 2007). Farré *et al.* (2009b) have discussed electrochemical immunosensors for environmental analysis (atrazine determination) which have used recombinant single-chain A_b (scA_b) fragments (Grennan *et al.*, 2003). Automated optical immunosensors have been used to detect many organic pollutants including estrone, progesterone, and testosterone in water samples. The detection limits were reported to be sub-ng/L (Taranova *et al.*, 2004). Labeled immunosensors have been used to detect hormones, enzymes, virus, tumor antigens, and bacterial antigens at concentrations around 10^{-12} to 10^{-9} mol/L (Campàs *et al.*, 2008; Wang *et al.*, 2008; Wang and Lin, 2008; Bojorge Ramírez *et al.*, 2009; He *et al.*, 2009)

Farré *et al.* (2007) list a number of limitations of the immunosensing approach including

(1) lengthy preparation time for immuno-reagents;
(2) lack of specificity as well as cross-reactivity;
(3) lack or limited response towards some groups of pollutants (e.g., perfluorinated compounds);
(4) poor stability under different thermal and pH conditions; and
(5) short life-times of biological components.

Farré *et al.* (2007) also enumerate key aspects that need to be addressed in future immunoassays which include

(1) development of more stable biological components;
(2) fabrication of more robust assays;
(3) assurance of better repeatability between different batches of production when disposable elements are involved; and
(4) integration of new technologies coupled to biosensors (e.g., the polymerase chain reaction).

González-Martinez *et al.* (2007) expect that immunosensors should be usable when

- a high number of samples need to be screened;
- on-line control is necessary;
- analysis is to be carried out in the field;
- different analytes need to be determined in a sample by different methods;
- data should be presented within minutes or in real time;
- samples need to be analyzed directly with none or hardly any pretreatment; and
- traditional methods do not work properly.

González-Martinez *et al.* (2007) described features of an ideal immunosensor as (1) very high sensitivity even to measure contaminants of interest in very diluted solutions; (2) high selectivity for compounds of interest without or with the minimum cross-reactivity problem; (3) applicable to the whole family of related compounds with generic immunoreagents; (4) high rapidity or speed without compromising sensitivity; (5) re-usable such that the device can work for a very long time (or large number of samples) without major maintenance; (6) capable of multiparameter determination (5–10 contaminants simultaneously); (7) versatile such that the sensor can be used for new analytes provided that the appropriate reagents are available; and (8) robust such that a sensor can be used under different conditions. Bojorge Ramírez *et al.* (2009) agree that there are a number of challenges to overcome for the mass production of immunosensors for a wide variety of compounds of interest. Antibodies to be used *in-vivo* need protein stability and antigen affinity. The industrial-scale mass production of antibodies is not yet possible without further advances in biochemical engineering technologies (Bojorge Ramírez *et al.,* 2009).

Fluorophores are now preferred over enzymes in immunosensors because they are more stable in solution than enzymes, also, the assay is shortened as the signal is displayed immediately (González-Martinez *et al.,* 2007). Wikipedia defines a fluorophore as a functional group of a molecule which becomes fluorescent under appropriate environmental conditions. Fluorophores absorb the energy of a specific wavelength and re-emit it at a different, but specific, wavelength. The quantity and wavelength of the emitted energy depend on both the fluorophore and the chemical environment to which the fluorophore is exposed (Joseph, 2006). Eu(III) chelate-dyed nanoparticles have been used as an antibody label in a fluoroimmunosensor for atrazine. The sensitivity (IC50) was reported to be ~1 μg L^{-1} for this immunosensor (Cummins *et al.*, 2006).

Nanotechnology has made significant inroads into the immunosensor area too. Magnetic nanoparticles were functionalized with specific antibodies (A_b) and used in immunomagnetic electrochemical sensors (Andreescu *et al.*, 2009). The use of A_b-coated magnetic nanoparticles eliminates or at least reduces the need for regeneration of the sensing surface. Quantification of the formed immunocomplex is done through enzyme labeling. Quantification can also be achieved via electrochemical detection of the reaction products after the complex is exposed to the enzymatic substrates, or through fluorescent labeling (Andreescu *et al.*, 2009). Different environmental contaminants were detected in this way. The contaminants include PCBs (Centi *et al.*, 2005), 2,4-dichlorophenoxy acetic acid based herbicide (also known as 2,4-D), and atrazine (Helali *et al.*, 2006; Zacco *et al.*, 2006). Arochlor 1248 (a PCB) detection limits of 0.4 ng/mL using screen-printed electrode strips have been achieved as well as atrazine detection limits of 0.027 nmol L^{-1} using anti-atrazine-specific antibody (Zacco *et al.*, 2006). Andreescu *et al.* (2009) reported in their review paper that paraoxon was measured at a low 12 μg/L level and a linear range within 24–1920 μg/L was achieved with an electrochemical immunosensor based on A_b-labeled gold nanoparticles on a glassy carbon electrode. Polymeric nano-particles (e.g., 2-methacryloyloxyethyl phosphorylcholine and polysterene) coated with anti-bisphenol A_b were used in a piezoelectric immunosensor for bisphenol A and an eight-fold sensitivity increase was achieved (Park *et al.*, 2006). In a reported work, Blăzkova *et al.* (2009) developed a simple and rapid immunochromatographic assay for a sensitive yet inexpensive monitoring of methiocarb in surface water using a binding inhibition format on a membrane strip. In the assay, the detection reagent consisted of anti-methiocarb A_b and colloidal carbon-labeled secondary A_b. They used carbon nanoparticles to bind proteins noncovalently without changing their bioactivity. A detection limit of 0.5 ng/L was reported. The assay results (recovery 90–106%) were in a good agreement with those of ELISA (recovery 91–117%). The strips were stable for at least 2 months without any change in performance. The developed immuno-chromatographic assay has potential for on-site screening of environmental contaminants.

A few examples of EDC sensors and biosensors are given in Table 3.1.

Table 3.1 Examples of EDC sensors and biosensors (after Rodriguez-Mozaz *et al.*, 2006; Farré *et al.*, 2009a, 2007)

Analyte	Transduction method	Limit of detection	Reference
Carbamates	Potentiometric	15–25 μM	Ivanov *et al.*, 2000
Dimethyl and diethyl dithiocarbamates	Amperometric detection	20 μM	Pita *et al.*, 1997
Bisphenol A	Potentiometric immunosensor	0.6 ng/mL	Mita *et al.*, 2007
Fenitrothion and ethyl p-nitrophenol	Organophosphates	4 μg/L	Rajasekar *et al.*, 2000
Progesterone	Amperometric detection	0.43 ng/mL	Carralero *et al.*, 2007
Parathion	Amperometric detection	10 ng/mL	Sacks *et al.*, 2000
2,4-dichlorophenoxy acetic acid	Amperometric immunosensor	0.1 μg/L	Wilmer *et al.*, 1997
Atrazine	Electrochemical amperometric	0.03 nmol/L	Zacco *et al.*, 2006
Chlorsulfuron	Electrochemical and amperometric	0.01 ng/mL	Dzantiev *et al.*, 2004
Estrogens	Total internal reflection fluorescence	0.05–0.15 ng/mL	Rodriguez-Mozaz *et al.*, 2006b
Trifluralin	Optical wave light spectroscopy	0.03 pg/mL	Székács *et al.*, 2003
Sulphamethoxazole	Piezoelectric	0.15 ng/mL	Melikhova *et al.*, 2006
Isoproturon	Total internal reflection fluorescence	0.01–0.14 μg/L	Blǎzkova *et al.*, 2006
Dioxins	Quartz crystal microbalances	15 ng/L	Kurosawa *et al.*, 2005
Paraoxon and carbofuran	Electrochemical (amperometric)	0.2 μg/L	Bachmann and Schmid, 1999
Phenols	Electrochemical	0.8 μg/L	Nistor *et al.*, 2002
Chlorophenols	Optical chemiluminescence	1.4–1975 μg/L	Degiuli and Blum, 2000
Nonylphenol	Electrochemical	10 μg/L	Evtugyn *et al.*, 2006

3.3 TRENDS IN SENSORS AND BIOSENSORS

3.3.1 Screen printed sensors and biosensors

There have been efforts to develop and apply environmentally friendly analytical procedures for contaminant detection. Increasing concerns about the impact of chemical waste generated during conventional analytical procedures have accelerated the search for alternatives. "Green" analytical chemistry is especially relevant to the use of instruments in the field and in decentralized laboratories where treatments for toxic and hazardous wastes are not available. A range of environmentally friendly electrochemical sensors for water monitoring is available and a number of them are now in advanced prototype stages. Conventional electrochemical cells are now being replaced with screen printed electrodes (SPEs) connected to miniaturized potentiostats. SPEs are finding uses in major analytical laboratory equipment and as hand-held field devices. A number of SPEs are commercially available and it is possible to manufacture them in the laboratory for research applications since screen-printing technology is getting cheaper and more easily available (Rico *et al.,* 2009). Farré *et al.* (2009b) reviewed many recent papers and covered many topics including SPEs. Electrochemical DNA and protein sensors based on the catalytic activity of hydrazine have been developed as SPEs (Shiddiky *et al.,* 2008). Enzyme-based high sensitivity biosensors for 2, 4-D, atrazine, and ziram have also been reported (Kim *et al.,* 2008). Gold nanoparticles have been used on tyrosinase electrode for the measurement of pesticides in water (Kim *et al.,* 2008). Organophosphorous and carbamate pesticides (e.g., monocrotophos, malathion, metasystox, and lannate) were measured electrochemically using SPEs containing immobilized acetylcholine esterase (AChE) (Dutta *et al.,* 2008). The measured concentration ranged from 0 to 10 μg/L (Farré *et al.,* 2009b).

3.3.2 Nanotechnology applications

Nanomaterials are natural or engineered materials which have at least one dimension at the nanometer scale ($\leq$100 nm). Nanomaterials possess completely new and enhanced properties as compared to the parent bulk materials. Examples of advanced nanomaterials include metallic, metal oxide, polymeric, semiconductor and ceramic nanoparticles, nanowires, nanotubes, quantum dots, nanorods, and composites of these materials. The unique properties of these materials are attributed to the extremely high surface area per unit weight, and their mechanical, electrical, optical, and catalytic properties. These properties offer a wide range of opportunities for the detection of environmental contaminants and toxins in addition to their remediation (Zhang, 2003; Li *et al.,* 2006;

Jimenez-Cadena *et al.*, 2007; Pillay and Ozoemena, 2007; Vaseashta *et al.*, 2007; Khan and Dhayal, 2008; Thompson and Bezbaruah, 2008; Bezbaruah *et al.*, 2009a,b). Nanotechnology incorporation into sensors (Trojanowicz, 2006; Ambrosi *et al.*, 2008; Gomez *et al.*, 2008; Guo and Dong, 2008; Kerman *et al.*, 2008; Wang and Lin, 2008; Algar *et al.*, 2009), miniaturization of electronics, and advancements in wireless communication technology have shown the emerging trend towards environmental sensor networks that continuously and remotely monitor environmental parameters (Huang *et al.*, 2001; Burda *et al.*, 2005; Liu *et al.*, 2005b; Jun *et al.*, 2006; Blasco and Picó, 2009; Zhang *et al.*, 2009).

Major research efforts are being targeted towards the development and application of nanomaterials in sensing (He and Toh, 2006; González-Martinez *et al.*, 2007). Nanomaterials are used in designing novel sensing systems and enhancing their performance (Farré *et al.*, 2009a). Satisfactory electrical communication between the active site of the enzyme and the electrode surface is a major challenge in amperometric enzyme electrodes. Aligned CNTs have been reported to improve electrical communication in such electrodes (Farré *et al.*, 2009a). Andreescu *et al.* (2009) have discussed environmental monitoring possibilities using nanomaterials. The paper cites a number of examples where nanotechnology has been successfully used in sensors. The use of nanotechnology in sensors and sensor hardware has resulted in the development of a number of miniaturized, rapid, ultrasensitive, and inexpensive methods for in-situ and field-based environmental monitoring. While these methods are not perfect and do not necessarily meet the general expectations, they are the harbingers of things to come. Nanoscale materials have been used for the construction of gas sensors (Gouma *et al.*, 2006; Jimenez-Cadena *et al.*, 2007; Milson *et al.*, 2007; Pillay and Ozoemena, 2007; Vaseashta *et al.*, 2007) enzyme sensors, immunosensors, and genosensors, for direct wiring of enzymes at electrode surfaces and for signal amplification (Liu and Lin, 2007; Pumera *et al.*, 2007). Metal oxide nanoparticles, which have excellent catalytic properties, are used for the construction of enzymeless electrochemical sensors (Hrbac *et al.*, 1997; Yao *et al.*, 2006; Hermanek *et al.*, 2007; Salimi *et al.*, 2007). Magnetic iron oxide nanostructures have potential for providing control for electrochemical processes (Wang *et al.*, 2005, 2006). Attachment of biological recognition elements to nanomaterial surfaces have led to the development of various catalytic and affinity biosensors (Andreescu *et al.*, 2009).

Costa-Fernández *et al.* (2006) have published a review on the application of quantum dots (QDs) as nanoprobes in sensing and biosensing. Andreescu *et al.* (2009) also discussed the use of carbon nanotubes (CNTs) and QDs in sensing. The surface chemistry, high surface area, and electronic properties make the use of CNTs ideal for chemical and biochemical sensing. CNTs have the ability to enhance the binding of biomolecules and increase electrocatalytic activities.

With CNTs, the detection of several analytes (e.g., pesticides) is possible at a low-applied potential. There is no need to use electronic mediators, and, hence, interferences are reduced. Electrochemical biosensors were constructed by immobilizing enzymes like acetylcholinesterase (AChE) and organophosphate hydrolyse (OPH) (Deo *et al.*, 2005) within CNTs/hybrid composites (Arribas *et al.*, 2005, 2007; Sha *et al.*, 2006; Rivas *et al.*, 2007). A number of strategies for CNT functionalization and application in sensing and biosensing have been reported (Andreescu *et al.*, 2005, 2008). These sensors have superior sensitivity compared to macroscale material-based ones. CNT-based sensors have been used for the monitoring of organophosphorus pesticides (Deo *et al.*, 2005; Joshi *et al.*, 2005), phenolic compounds (Sha *et al.*, 2006), and herbicides (Arribas *et al.*, 2005, 2007). Gold, platinum, copper, and some other nanoparticles have been incorporated onto CNTs/polymeric composites to further enhance their characteristics (Andreescu *et al.*, 2009). Further, signal amplification has been achieved using gold nanoparticle in biosensors and in a variety of colorimetric and fluorescence assays (Andreescu *et al.*, 2009). QDs have been used as sensing probes for small metal ions (Costa-Fernández *et al.*, 2006; Somers *et al.*, 2007), pesticides (Ji *et al.*, 2005), phenols (Yuan *et al.*, 2008), and nitroaromatic explosives (Goldman *et al.*, 2005). QD-enzyme conjugates respond to enzyme substrates and inhibitors (Ji *et al.*, 2005), and antibodies (Goldman *et al.*, 2005). The conjugates approach has been used to fabricate QDs (Abad *et al.*, 2005; Ji *et al.*, 2005) for the detection of pesticides. Photoluminescence intensity of the QD bioconjugates changes in the presence of the analyte (e.g., pesticide paraoxon) and the changes in signal have been quantified and correlated to analyte concentrations (Ji *et al.*, 2005).

3.3.3 Molecular imprinted polymer sensors

A conventional biosensor selectively recognizes analytes and binds them into the specific binding layers provided. This binding creates different events such as optical, mass, thermal and electrochemical changes that produce their corresponding signals (Eggins, 2002). Lots of progress has been made on biological recognition, but there are still some complex compounds that can't be detected accurately by biosensors. Such compounds include antibodies and enzymes (Sellergren, 2001; Yan and Ramstrom, 2005). Molecularly imprinted polymers (MIPs) have drawn considerable attention in recent years. In the environmental area, they are being used for remediation and sensing. MIPs are synthesized using template (target) molecules which are cross-linked into a monomer. The cross linkers are specific to the template. The target-monomer complex is then polymerized and the template molecules are removed to leave the polymer matrix with "holes" specific to the target molecules (Haupt and

Mosbach, 2000; Widstrand *et al.*, 2006). The MIP synthesis process is illustrated in Figure 3.1. The holes capture the target molecules from a sample even if they are present in small amounts. MIP materials have high recognition affinity to the target molecule. The holes are so specific that they allow only the target molecules to enter them and reject all others. MIPs are robust, cost effective, and easy to design. There are a number of advantages of MIP materials: (1) they are small in size; (2) they have increased number of accessible complementary cavities for the target molecule; (3) they have enhanced surface catalytic activity; and (4) they can establish equilibration with the target molecule very quickly due to the limited diffusional length (Nakao and Kaeriyama, 1989; Lu *et al.*, 1999). MIPs have been used in conjunction with optical and electrochemical techniques to detect amino acids, enzymes, antibodies, pesticides, proteins, and vitamins.

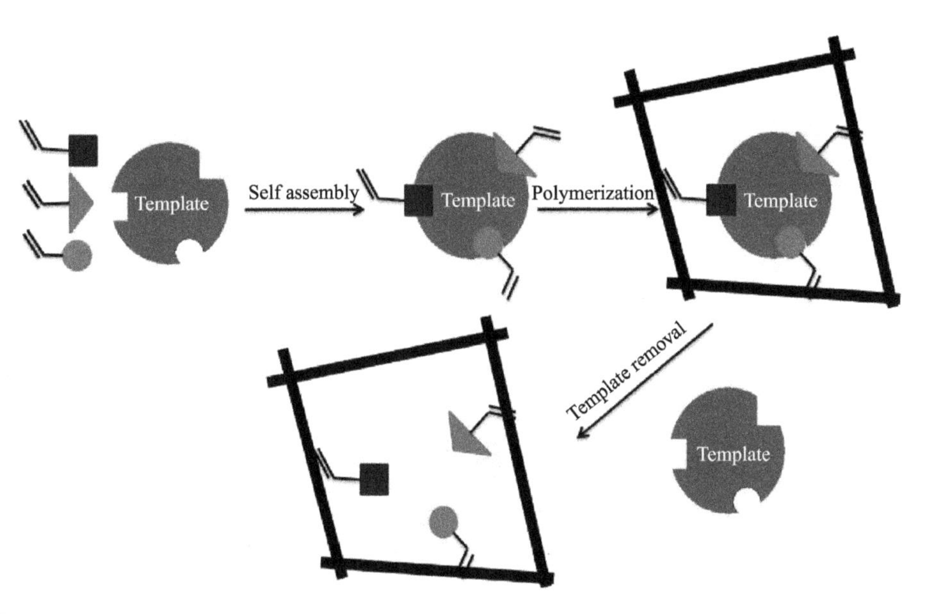

Figure 3.1 Schematic representation of molecular imprinted polymer (after Shelke *et al.*, 2008)

The efficiency of the MIP sensors depends on the interaction of the template molecule and the complementary functional monomer group (Whitcombe and Vulfson, 2001). Noncovalent and covalent interactions are observed between them. The noncovalent interactions that hold the template molecule and the functional monomer group together include hydrogen bonding, hydrophobic interaction, Van der Waals forces, and dipole-dipole interactions (Holthoff and Bright, 2007). However, if a functional group has strong covalent interactions, nonconvalent interactions are suppressed (Graham *et al.,* 2002). Reversible covalent interactions can also bind the template molecule with the functional monomer group. Wulff *et al.* (1977) first introduced the concept of covalent interactions between the functional group and the template molecule and how the template molecule can be released by cleaving the bond. These types of interactions are favored if the functional monomer has diol, aldehyde, or amine. A MIP sensor selectively binds the analyte molecules and produces a transduction scheme to detect the analyte (Figure 3.2) (Lange *et al.,* 2008). A few examples of MIP-based sensors are listed in Table 3.2. The list includes the template molecules, transduction method, and detection limits for various analytes. Both organic and inorganic materials are used to synthesize the MIPs used in the sensors. Holthoff and Bright (2007) have discussed the use of polystyrene, polyacrylate, and inorganic polysiloxane to synthesize MIP-based sensors.

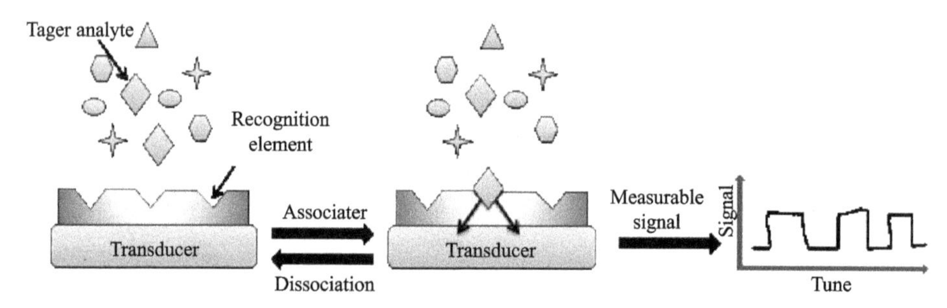

Figure 3.2 Schematic Representation of an MIP-based biosensor and its response profile (after Holthoff and Bright 2007)

Conventional MIP-based sensors are typically designed for individual analytes. The use of nanotechnology in molecular imprinting holds a lot of promise to overcome this limitation. Micro- and nano-sensors are of interest to scientists and engineers as they can be used as arrays to analyze different molecules at the same time (Alexander *et al.,* 2006). MIP materials can be appropriately patterned on a chip surface and fabricated for multi-analyte sensing applications by interfacing with transducers. Various patterning techniques that are used to make these MIP-based micro- and nano-sensors include photo-lithography, soft-lithography, and microspotting. UV mask lithography has been used to make the micro- and nano-MIP sensors where MIP layer is applied onto a metallic electrode and then cured by UV irradiation (Huang *et al.,* 2004). For example, Pt electrode is used in acrylic molecularly imprinted photoresist (Du *et al.,* 2008; Gomez-Caballero *et al.,* 2008), and Au and Pt electrodes are used to fabricate albuterol (a bronchodilator) MIP microsensor (Huang *et al.,* 2007).

Table 3.2 Examples of MIP-based sensors (after Holthoff and Bright 2007; Navarro-Villoslada *et al.*, 2007; Zhang *et al.*, 2008)

Target analyte	Template	Transduction method	Detection limit	Reference
Atrazine	Atrazine	Electrochemical	0.5 μM	Prasad *et al.*, 2007
Cytidine	Cytidine	Electrochemical	Not reported	Whitcombe *et al.*, 1995
Glutathione	Glutathione	Electrochemical	1.25 μM	Yang *et al.*, 2005
L-Histidine	LHistidine	Electrochemical	25 nM	Zhang *et al.*, 2005
Parathion	Parathion	Electrochemical	1 nM	Li *et al.*, 2005
L-Tryptophan	L-Tryptophan	Optical	Not reported	Liao *et al.*, 1999
Adrenaline	Adrenaline	Optical	5 μM	Matsui *et al.*, 2004
1,10-Phenanthroline	1,10-Phenanthroline	Optical	Not reported	Lin and Yamada, 2001
9-Ethyladenine	9-Ethyladenine	Optical	Not reported	Matsui *et al.*, 2000
9-Anthrol	9,10-Anthracenediol	Optical	0.3 μM	Shughart *et al.*, 2006
2,4-Dichlorophenoxy-acetic acid	2,4-Dichlorophenoxy-acetic acid	Optical	Not reported	Leung *et al.*, 2001
Penicillin G	Penicillin G	Optical	1 ppm	Zhang *et al.*, 2008
Zearalenone	Zearalenone	Optical	25 μM	Navarro-Villoslada *et al.*, 2007

Microcontact printing is one of the emerging technologies for the production of patterned microstructures (Quist *et al.*, 2005; Lin *et al.*, 2006). MIP micropatterns are created using this microcontact printing method. MIP microstructures can be synthesized using the poly(dimethyl-siloxiane) (PDMS) stamp technique (Yan and Kapua, 2001). However, incompatibility of the PDMS stamps to some organic solvents limits their applications (Vandevelde *et al.*, 2007). MIP based sensor for theophylline (a methylxanthine drug, also known as dimethylxanthine) has been synthesized using this technique and excellent selectivity for the template molecule was achieved. Similar results were reported for structurally similar caffeine (Voicu *et al.*, 2007). Micro-stereolithography technique is also used to synthesize MIP based sensors with 9-ethyl adenine as the template (Conrad *et al.*, 2003).

3.3.4 Conducting polymers

Conducting polymers have found increased applications in various industries. Some of the main classes of conducting polymers that are available for various applications include polyacetylene, polyaniline (PANI), polypyrrole (PPY), polythiophene (PTH), poly(paraphenylene), poly(paraphenylenevinylene), polyfluorene, polycarbazole, and polyindole (PI). Conducting polymers exhibit intrinsic conductivity when the conjugated backbone of the polymer is oxidized or reduced (Bredas, 1995). Apart from its conductivity, the change of electronic band in the conducting polymer affects the optical properties in the UV-visible and near IR region. The changes in conductivity and optical properties make them candidates for use as optical sensors. Chemical and electrochemical methods are used to inject charge (doping) into conducting polymers (Wallace 2003). The electrochemical method is preferred as it is easy to adjust the doping level by controlling the electrical potential.

Conducting polymers have been effectively used to detect metal ions. Polyindole and polycarbazole provide selective responses towards Cu(II) ions (Prakash *et al.,* 2002) and poly-3-octylthiophene (P3OTH) gives Nernstian responses to Ag(I) ions (Vazquez *et al.,* 2005). Extraction and stripping voltammetry method has also been used to detect Pb(II) and Hg(II) with conductive polymers (Heitzmann *et al.,* 2007). Further, ion selectivity of the sensors can be improved by introducing specific ligands (Migdalski *et al.,* 2003; Zanganeh and Amini, 2007; Mousavi *et al.,* 2008), ionophores (Cortina-Puig *et al.,* 2007) and monomers (Seol *et al.,* 2004; Heitzmann *et al.,* 2007) to the polymer backbone.

Organic molecules have an affinity towards the conducting polymer backbone, side group, and to the immobilized receptor group. This affinity is exploited to design conducting polymer sensors for organic molecules. Both biological and synthetic receptors can be used to selectively bind the organic molecule. Dopamine, ascorbic acid, and chlorpromazine sensing has been done by introducing γ-cyclodextrin receptor to poly(3-methylthiophene) (P3MTH) (Bouchta *et al.,* 2005) and β-cyclodextrin to PPY (Izaoumen *et al.,* 2005). A film of PANI and poly(3-aminophenylboronic acid) is used for the detection of saccharides by optical method (Pringsheim *et al.,* 1999). Syntheses of a variety of chemosensitive PANI and PPY conductive polymers to detect dicarboxylates, amino acids and ascorbic acid have also been reported (Volf *et al.,* 2002).

The electropolymerization method is also used to synthesize MIPs for the preparation of conductive chemosensitive film (Gomez-Caballero *et al.,* 2005; Yu *et al.,* 2005; Liu *et al.,* 2006). The electropolymerization method can control the thickness of the polymer film and this technique is compatible with the combinatorial and high-throughput approach (Potyrailo and Mirsky, 2008). Conducting MIP film of PANI has been synthesized to detect ATP, ADP, and AMP (Sreenivasan, 2007).

Synthetic and biological receptors can be used to manipulate the sensitivity of a conducting polymer for different analytes (Adhikari and Majunder, 2004; Ahuja *et al.*, 2007). Some conducting polymers that have been modified with various receptors are listed in Table 3.3. To immobilize the receptor, it is bonded to the polymer matrix through covalent or noncovalent interaction. Physical adsorption (Lopéz *et al.*, 2006), the Langmuir-Blodgett technique (Sharma *et al.*, 2004), layer-by layer deposition technique (Portnov *et al.*, 2006), and mechanical embedding method (Kan *et al.*, 2004) are used to bind the receptor to the matrix through noncovalent bonding. Gerard *et al.* (2002) have discussed the advantages and limitations of these techniques.

Table 3.3 Conducting polymer-based sensors and biosensors (after Lange *et al.*, 2008)

Analyte	Receptor	Polymer	Transduction method	Reference
Uric acid	Uricase	PANI	Optical, Amperometric	Arora *et al.*, 2007; Kan *et al.*, 2004
H_2O_2	Horseradish peroxidase	PANI/ polyethylene terephthalate	Optical	Caramori and Fernandes, 2004; Borole *et al.*, 2005; Fernandes *et al.*, 2005
Glucose	Glucose oxidase	3-methylthio-phene/thio-phene-3-acetic acid copolymer	Amperometric	Kuwahara *et al.*, 2005
Phenol	Tyrosinase	Polyethylene-dioxythiophene	Amperometric	Vedrine *et al.*, 2003
Organipho-sphate pesticide	Acetylcholine-sterase	PANI	Amperometric	Law and Higson, 2005
Cholesterol	Cholesterol esterase/ Cholesterol oxidase	PPY, PANI	Amperometric	Singh *et al.*, 2004; Singh *et al.*, 2006
Glycoproteins	Boronic acid	Poly(aniline boronic acid)	Optical	Liu *et al.*, 2006

3.4 FUTURE OF SENSING

Sensors and biosensors have a number of disadvantages compared to standard chemical monitoring methods, however, they fulfill a number of requirements of current and emerging environmental pollution monitoring that chemical methods fail to address. Ongoing developments in material technology, computer technology, and microelectronics are expected to help sensor developers to overcome many of these problems. It is expected that progress in the development of tools and strategies to identify, record, store, and transmit parameter data will help in expanding the scope of the use of sensors on a broader scale (Blasco and Picó, 2009).

Additionally, the next generation of environmental sensors should operate as stand-alone outside the laboratory environment and with remote controls. New devices based on microelectronics and related (bio)-micro-electro-mechanical systems (MEMS) and (bio)-nano-electro-mechanical systems (NEMS) are expected to provide technological solutions. Miniaturized sensing devices, microfluidic delivery systems, and multiple sensors on one chip are needed. High reliability, potential for mass production, low cost of production, and low energy consumption are also expected and some progress has already been achieved in these areas (Farré *et al.*, 2007).

The recent developments in communication technology have not yet been fully exploited in the sensor area. New technologies like Bluetooth, WiFi and radio-frequency identification (RFID) can definitely be utilized to provide a network of distributed electronic devices in even very remote places. A wireless sensor network comprising spatially-distributed sensors or biosensors to monitor environmental conditions will contribute enormously towards continuous environmental monitoring especially in environments that are currently difficult to monitor such as coastal areas and open seas (Farré *et al.*, 2009a). Blasco and Picó (2009) expect that such a network can: (1) provide appropriate feedback during characterization or remediation of contaminated sites; (2) offer rapid warning in the case of sudden contamination; and (3) minimize the huge labor and analytical costs, as well as errors and delays, inherent to laboratory-based analyses. The laboratory-on-a-chip (LOC) is another concept that is going to impact future sensor technology. LOC involves microfabrication to achieve miniaturization and/or minimization of components of the analytical process (sample preparation, hardware, reaction time and detection) (Farré *et al.*, 2007). It has been suggested that nanoscale and ultra-miniaturized sensors could dominate the production lines in the next generation of biotechnology-based industries (Farré *et al.*, 2007).

3.5 REFERENCES

Abad, J. M., Mertens, S. F. L., Pita, M. Fernandez, V. F. and Schiffrin, D. J. (2005) Functionalization of Thioctic acid-capped gold nanoparticles for specific immobilization of histidine-tagged proteins. *J. Am. Chem. Soc.* **127**(15), 5689–5694.

Adhikari, B. Majumdar, S. (2004) Polymers in sensor applications. *Prog. Polym. Sci.* **29**(7), 699–766.

Ahuja, T., Mir, I. A., Kumar, D. and Rajesh (2007) Biomolecular immobilization on conducting polymers for biosensing applications. *Biomaterials* **28**(5), 791–895.

Alexander, C., Andersson, H. S., Andersson, L. I., Ansell, R. J., Kirsch, N., Nicholls, I. A., O'Mahony J. and Whitcombe, M. J. (2006) Molecular imprinting science and technology: A survey of the literature for the years up to and including 2003. *J. Mol. Recognit.* **19**(2), 106–180.

Algar, W. R., Massey, M. and Krull, U. J. (2009) The application of quantum dots, gold nanoparticles and molecular switches to optical nucleic-acid diagnostics. *Trends Anal. Chem.* **28**(3), 292–306.

Allan, I. J., Vrana, B., Greenwood, R., Mills, D. W., Roig, B. and Gonzalez, C. (2006) A "toolbox" for biological and chemical monitoring requirements for the European Union's Water Framework Directive. *Talanta* **69**(2), 302–322.

Ambrosi, A., Merkori, A. and de la Escosura-Muniz, A. (2008) Electrochemical analysis with nanoparticle-based biosystems. *Trends Anal. Chem.* **27**(7), 568–584.

Arora, K., Sumana, G., Saxena, V., Gupta, R. K., Gupta, S. K., Yakhmi, J. V., Pandey, M. K., Chand, S. and Malhotra, B. D. (2007) Improved performance of polyaniline-uricase biosensor. *Acta* **594**(1), 17–23.

Asano, K., Ono, A., Hashimoto, S., Inoue, T. and Kanno, J. (2004) Screening of endocrine disrupting chemicals using a surface plasmon resonance sensor. *Anal. Sci.* **20**(4), 611–616.

Andreescu, D., Andreescu, S. and Sadik, O. A. (2005) New materials for biosensors, biochips and molecular bioelectronics. In: *Biosensors and Modern Biospecific Analytical Techniques*, (ed. Gorton, L.), Elsevier, Amsterdam, pp. 285–327.

Andreescu, S., Njagi, J., Ispas, C. and Ravalli. M. T. (2009) JEM spotlight: Applications of advanced nanomaterials for environmental monitoring. *J. Environ. Monitor.* **11**(1), 27–40.

Andreescu, S., Njagi, J. Ispas, C. (2008) Nanostructured materials for enzyme immobilization and biosensors. In: *The New Frontiers of organic and composite nanotechnology*, (eds Erokhin, V., Ram, M. K. and Yavuz, O.), Elsevier, Amsterdam.

Arribas, A. S., Bermejo, E., Chicharro, M., Zapardiel, A., Luque, G. L., Ferreyra, N. F. amd Rivas, G. A. (2007) Analytical applications of glassy carbon electrodes modified with multi-wall carbon nanotubes dispersed in polyethylenimine as detectors in flow systems. *Anal. Chim. Acta* **596**(2), 183–194.

Arribas, A. S., Vazquez, T., Wang, J., Mulchandani, A. and Chen, W. (2005) Electrochemical and optical bioassays of nerve agents based on the organophosphorus-hydrolase mediated growth of cupric ferrocyanide nanoparticles. *Electrochem. Commun.* **7**(12), 1371–1374.

Bachmann, T. T. and Schmid, R. D. (1999) A disposable multielectrode biosensor for rapid simultaneous detection of the insecticides paraoxon and carbofuran at high resolution. *Anal. Chim. Acta* **401**(1–2), 95–103.

Barcelo, D. and Petrovic, M. (2006) New concepts in chemical and biological monitoring of priority and emerging pollutants in water. *Anal. Bioanal. Chem.* **385**(6), 983–984.

Berlein, S., Spener, F. and Zaborosch, C. (2002) Microbial and cytoplasmic membrane-based potentiometric biosensors for direct determination of organophosphorus insecticides. *Appl. Microbiol. Biotechnol.* **54**(5), 652–658.

Bezbaruah, A. N., Krajangpan, S., Chisholm, B. J., Khan, E. and Elorza Bermudez, J. J. (2009a) Entrapment of iron nanoparticles in calcium alginate beads for groundwater remediation applications, *J. Hazard. Mater.* **166**(2–3), 1339–1343.

Bezbaruah, A. N., Thompson, J. M. and Chisholm, B. J. (2009b) Remediation of alachlor and atrazine contaminated water with zero-valent iron nanoparticle, *J. Environ. Sci. Heal. B* **44**(6), 1–7.

Blasco, C. and Picó, Y. (2009) Prospects for combining chemical and biological methods for integrated environmental assessment. *Trends Anal. Chem.* **28**(6), 745–757.

Blăžkova, M., Mickova-Holubova, B., Rauch, P. and Fukal. L. (2009) Immunochromatographic colloidal carbon-based assay for detection of methiocarb in surface water. *Biosensors and Bioelectronics* **25**(4), 753–758.

Blăžkova, M., Karamonova, L., Greifova, M., Fukal, L., Hoza, I., Rauch, P. and Wyatt, G. (2006) Development of a rapid, simple paddle-style dipstick dye immunoassay specific for Listeria monocytogenes. *Eur. Food Res. Technol.* **223**(6), 821–827.

Bojorge Ramirez, N., Salgado, A. M. and Valdman, B. (2009) The evolution and development of immunosensors for health and environmental monitoring: Problems and perspectives, *Braz. J. Chem. Eng.* **26**(2), 227–249.

Borole, D. D., Kapadi, U.R, Mahulikar, P. P. and Hundiwale, D. G. (2005) Glucose oxidase electrodes of a terpolymer poly(aniline-co-o-anisidine-co-o-toluidine) as biosensors. *Eur. Polym. J.* **41**(9), 2183–2188.

Bouchta, D., Izaoumen, N., Zejli, H., Kaoutit, M. E. and Temsamani, K. R. (2005) A novel electrochemical synthesis of poly-3-methylthiophene-gamma-cyclodextrin film – Application for the analysis of chlorpromazine and some neurotransmitters. *Biosens. Bioelectron.* **20**(11), 2228–2235.

Brack, W., Klamer, H. J. C., de Ada, M. L. and Barcelo, D. (2007) Effect-directed analysis of key toxicants in European river basins – A review. *Environ. Sci. Pollut. Res.* **14**(1), 30–38.

Bredas, J. L. and Street, G. B. (1985) Polarons, bipolarons, and solitons in conducting polymers. *Acc. Chem. Res.* **18**(10), 309–315.

Burda, C. Chen, X., Narayanan, R. and EI-Sayed, M. A. (2005) Chemistry and Properties of Nanocrystals of Different Shapes. *Chem. Rev.* **105**(4), 1025–11-2.

Campàs, M., Carpentier, R. and Rouillon, R. (2008) Plant tissue-and photosynthesis-based biosensors. *Biotechnol. Adv.* **26**(4), 370–378.

Caramori, S. S. and Fernandes, K. F. (2004) Covalent immobilisation of horseradish peroxidase onto poly (ethylene terephthalate)-poly (aniline) composite. *Process Biochem.* **39**(7), 883–888.

Carralero, V., Gonzalez-Cortes, A., Yanez-Sedeno, P. and Pingarron, J. M. (2007) Nanostructured progesterone immunosensor using a tyrosinase-colloidal gold-graphite-Teflon biosensor as amperometric transducer. *Anal. Chim. Acta* **596**(1), 86–91.

Centi, S., Laschi, S., Franek, M. and Mascini, M. (2005) A disposable immunomagnetic electrochemical sensor based on functionalised magnetic beads and carbon-based screen-printed electrodes (SPCEs) for the detection of polychlorinated biphenyls (PCBs). *Anal. Chim. Acta*, **538**(1–2), 205–212.

Chaplen, F. W. R., Vissvesvaran, G., Henry, E. C. and Jovanovic, G. N. (2007) Improvement of bioactive compound classification through integration of orthogonal cell-based biosensing methods. *Sensors*, **7**(1), 38–51.

Comerton, A. M., Andrews, R. C. and Bagley, D. M. (2009) Practical overview of analytical methods for endocrine-disrupting compounds, pharmaceuticals and personal care products in water and wastewater. *Philosophical Transactions of Royal Society A*, **367**(1904), 3923–3939.

Conrad, P. G., Nishimura, P. T., Aherne, D., Schwartz, B. J., Wu, D., Fang, N., Zhang, X., Roberts, M. J. and Shea, K. J. (2003) Functional molecularly imprinted polymer microstructures fabricated using microstereolithographic techniques. *Adv. Mater.* **15**(18), 1541–1514.

Cortina-Puig, M., Munoz-Berbel, X., del Valle, M., Munoz, F. J. and Alonso-Lomillo, M. A. (2007) Characterization of an ion-selective polypyrrole coating and application to the joint determination of potassium, sodium and ammonium by electrochemical impedance spectroscopy and partial least squares method. *Anal. Chim. Acta* **597**(2), 231–237.

Costa-Fernández, J. M., Pereiro, R. and Sanz-Medel, A. (2006) The use of luminescent quantum dots for optical sensing.*Trends. Anal. Chem.* **25**(3), 207–218.

Cummins, C. M., Koivunen. M. E., Stephanian, A., Gee, S. J., Hammock, B. D. and Kennedy I. M. (2006) Application of europium(III) chelate-dyed nanoparticle labels in a competitive atrazine fluoroimmunoassay on an ITO waveguide. *Biosens Bioelectron.* **21**(7), 1077–1085.

Degiuli, A. and Blum, L. J. (2000) Flow injection chemiluminescence detection of chlorophenols with a fiber optic biosensor. *J. Med. Biochem.* **4**(1), 32–42.

Deo, R. P., Wang, J., Block, I., Mulchandani, A., Joshi, K. A., Trojanowicz, M., Scholz, F., Chen, W. and Lin, Y. (2005) Determination of organophosphate pesticides at a carbon nanotube/organophosphorus hydrolase electrochemical biosensor. *Anal. Chim. Acta.* **530**(2), 185–189.

Du, D., Chen, S., Cai, J., Tao, Y., Tu, H. and Zhang, A. (2008) Recognition of dimethoate carried by bi-layer electrodeposition of silver nanoparticles and imprinted poly-*o*-phenylenediamine. *Electrochim. Acta* **53**(22), 6589- 6595.

Dutta, K., Bhattacharyay, D., Mukherjee, A., Setford, S. J., Turner, A. P. F. and Sarkar, P. (2008) Detection of pesticide by polymeric enzyme electrodes. *Ecotoxicol. Environ. Safety* **69**(3), 556–561.

Dzantiev, B. B., Yazynena, E. V., Zherdev, A. V., Plekhanova, Y. V., Reshetilov, A. N., Chang, S.-C. and McNeil, C. J. (2004) Determination of the herbicide chlorsulfuron by amperometric sensor based on separation-free bienzyme immunoassay. *Sens. Actuators. B* **98**(2–3), 254–261.

Eggins, B. R. (ed.), (2002) *Chemical Sensors and Biosensors*, John Wiley & Sons, Chichester, UK.

Evtugyn, G. A., Eremin, S. A., Shaljamova, R. P., Ismagilova, A. R. and Budnikov, H. C. (2006) Amperometric immunosensor for nonylphenol determination based on peroxidase indicating reaction. *Biosens. Bioelectron.* **22**(1), 56–62

Le Blanc, F. A., Albrecht, C., Bonn, T., Fechner, P., Proll, G., Pröll, F., Carlquist, M. and Gauglitz, G. (2009) A novel analytical tool for quantification of estrogenicity in river water based on fluorescence labelled estrogen receptor α. *Anal. Bioanal. Chem.* **395**(6), 1769–1776.

Farré, M., Kantiani, L., Perez, S. and Barcelo. D. (2009a) Sensors and biosensors in support of EU Directives, *Trends Anal. Chem.* **28**(2), 170–185.

Farré, M. Gajda-Schrantz, K. Kantiani, L. and Barcelo, D. (2009b) Ecotoxicity and analysis of nanomaterials in the aquatic environment. *Anal. Bioanal. Chem.* **393**(1), 81–95.

Farré, M., Kantiani, L. and Barcelo, D. (2007) Advances in immunochemical technologies for analysis of organic pollutants in the environment, *Trends Anal. Chem.* **26**(11), 1100–1112.

Fechner, P., Proell, F., Carlquist, M. and Proll, G. (2009) An advanced biosensor for the prediction of estrogenic effects of endocrinedisrupting chemicals on the estrogen receptor alpha. *Anal. Bioanal. Chem.* **393**(6–7), 1579–1585.

Fernandes, K. F., Lima, C. S., Lopes, F. M. and Collins, C. H. (2005) Hydrogen peroxide detection system consisting of chemically immobilised peroxidase and spectrometer. *Process Biochem.* **40**(11), 3441–3445.

Fernandez, M. P., Noguerol, T. N., Lacorte, S., Buchanan, I. and Pina, B. (2009) Toxicity identification fractionation of environmental estrogens in waste water and sludge using gas and liquid chromatography coupled to mass spectrometry and recombinant yeast assay. *Anal. Bioanal. Chem.* **394**(3), 957–968.

Gatidou, G., Thomaidis, N., Stasinakis, A. and Lekkas, T. (2007) Simultaneous determination of the endocrine disrupting compounds nonylphenol, nonylphenol ethoxylates, triclosan and bisphenol A in wastewater and sewage sludge by gas chromatography-mass spectrometry. *J. Chromatogr. A* **1138**(1–2), 32–41.

Gawrys, M. D., Hartman, I. Landweber, L. F. and Wood. D. W. (2009) Use of engineered *Escherichia coli* cells to detect estrogenicity in everyday consumer products, *J. Chem. Technol. Biotechnol.* **84**(12), 1834–1840.

Gerard, M., Chaubey, A. and Malhotra, B. D. (2002) Application of conducting polymers to biosensors. *Biosens. Bioelectron.* **17**(5), 345–359.

Giardi, M. T., Scognamiglio, V., Rea, G., Rodio, G., Antonacci, A., Lambreva, M., Pezzotti, G. and Johanningmeier U. (2009) Optical biosensors for environmental monitoring based on computational and biotechnological tools for engineering the photosynthetic D1 protein of *Chlamydomonas reinhardtii. Biosensors and Bioelectronics* **25**(2), 294–300.

Giardi, M. T. and Pace, E. (2006) Photosystem II-Based Biosensors for the Detection of Photosynthetic Herbicides, Maria Teresa Giardi and Emanuela Pace. In: *Biotechnological Applications of Photosynthetic Proteins: Biochips, Biosensors and Biodevices* (eds. Giardi, M. T. and Piletska, E.), Landes Bioscience, Springer Publishers, Georgetown, TX, USA, pp. 147–154.

Giardi, M. T. and Pace, E. (2005) Photosynthetic proteins for technological applications. *Trends Biotechnol.* **23**(5), 257–263.

Goldman, E. R., Meditnz, I. L., Whitley, J. L., Hayhurst, A., Clapp, A. R., Uyenda, H. T., Deschamps, J. R., Lessman, M. E. and Mattoussi, H. (2005) A hybrid quantum dot-antibody fragment fluorescence resonance energy transfer-based TNT sensor. *J. Am. Chem. Soc.* **127**(18), 6744–6751.

Gomez-Caballero, A., Goicolea, M. A. and Barrio, R. J. (2005) Paracetamol voltammetric microsensors based on electrocopolymerized-molecularly imprinted film modified carbon fiber microelectrodes. *Analyst* **130**(7), 1012–1018.

Gomez-Caballero, A., Unceta, N., Aranzazu Goicolea, M. and Barrio, R. J. (2008) Evaluation of the selective detection of 4,6-dinitro-*o*-cresol by a molecularly imprinted polymer based microsensor electrosynthesized in a semiorganic media. *Sens. Actuators B* **130**(2), 713–722.

Gomez, M. J., Fernandez-Romero, J. M. and Guilar-Caballos, M. P. (2008) Nanostructures as analytical tools in bioassays. *Trends Anal. Chem.* **27**(5), 394–406.

Gonzalez-Doncel, M., Ortiz, J., Izquierdo, J. J., Martın, B., Sanchez, P. and Tarazona, J. V. (2006) Statistical evaluation of chronic toxicity data on aquatic organisms for the hazard identification: The chemicals toxicity distribution approach. *Chemosphere* **63**(5), 835–844.

González-Martinez M. A., Puchades R. and Maquieira, A. (2007) Optical immunosensors for environmental monitoring: How far have we come? *Anal. Bioanal. Chem.* **387**(1), 205–218.

Gouma, P. I., Prasad, A. K. Iyer, K. K. (2006) Selective nanoprobes for 'signalling gases'. *Nanotechnol.* **17**(4), S48–S53.

Graham, A. L., Carlson, C. A. and Edmiston, P. L. (2002) Development and characterization of molecularly imprinted Sol-Gel materials for the selective detection of DDT. *Anal. Chem.* **74**(2), 458–467.

Grennan, K., Strachan, G., Porter, A. J. Killard, A. J. and Smyth, M. R. (2003) Atrazine analysis using an amperometric immunosensor based on single-chain antibody fragments and regeneration-free multi-calibrant measurement. *Anal. Chem. Acta* **500**(1–2), 287–298.

Grote, M., Brack, W., Walter, H. A. and Altenburger, R. (2005) Confirmation of cause-effect relationships using effect-directed analysis for complex environmental samples. *Environ. Toxicol. Chem.* **24**(6), 1420–1427.

Guo, S. and Dong, S. (2008) Biomolecule-nanoparticle hybrids for electrochemical biosensors. *Trends Anal. Chem.* **28**(1), 96–109.

Haupt, K. and Mosbach, K. (2000) Molecularly imprinted polymers and their use in biomimetic sensors. *Chem. Rev.* **100**(7), 2495–2504.

He, L. and Toh C. S. (2006) Recent advances in anal chem-a material approach. *Anal.Chem. Acta* **556**(1), 1–15.

He, P., Wang, Z., Zhang, L. and Yang, W. (2009) Development of a label-free electrochemical immunosensor based on carbon nanotube for rapid determination of clenbuterol. *Food Chem.* **112**(3), 707–714.

Heitzmann, M., Bucher, C., Moutet, J. C., Pereira, E., Rivas, B. L., Royal, G. and Saint-Aman, E. (2007) Complexation of poly (pyrrole-EDTA like) film modified electrodes: Application to metal cations electroanalysis. *Electrochim. Acta* **52**(9), 3082–3087.

Helali, S., Martelet, C., Abdelghani, A., Maaref, M. A. and Jaffrezic-Renault, N. (2006) A disposable immunomagnetic electrochemical sensor based on functionalized magnetic beads on gold surface for the detection of atrazine. *Electrochim. Acta,* **51**(24), 5182–5186.

Hermanek, M., Zboril, R., Medrik, I., Pechousek, J. and Gregor, C. (2007) Catalytic efficiency of iron(III) oxides in decomposition of hydrogen peroxide: Competition between the surface area and crystallinity of nanoparticles *J. Am. Chem. Soc.* **129**(35), 10929–10936.

Holthoff, E. L. and Bright, E. V. (2007) Molecularly templated materials in chemical sensing. *Anal. Ciem. Acta.* **594**(2), 147–161.

Hrbac, J., Halouzka, V., Zboril, R., Papadopoulos, K. and Triantis, T. (1997) Carbon electrodes modified by nanoscopic iron(III) oxides to assemble chemical sensors for the hydrogen peroxide amperometric detection. *Analys.* **122**(17), 985–989.

Huang, H. C., Lin, C. I., Joseph, A. K. and Lee, Y. D. (2004) Photo-lithographically impregnated and molecularly imprinted polymer thin film for biosensor applications. *J. Chromatogr. A* **1027**(1–2) 263–268.

Huang, H. C., Huang, S. Y., Lin, C. I. and Lee, Y. D. (2007) A multi-array sensor via the integration of acrylic molecularly imprinted photoresists and ultramicroelectrodes on a glass chip. *Anal. Chim. Acta* **582**(1), 137–146.

Huang, Y., Duan, X., Wei, Q. and Liber, C. M. (2001) Directed assembly of one-dimensional nanostructures into functional networks. *Science* **291**(5504), 630–633.

Huertas-Perez, J. F. and Garcia-Campana, A. M., (2008) Determination of N-methylcarbamate pesticides in water and vegetable samples by HPLC with post-column chemiluminescence detection using the luminol reaction. *Anal. Chim. Acta* **630**(2), 194–204.

Ivanov, A. N., Evtugyn, G. A., Gyurcsanyi, R. E., Toth, K. and Budnikov, H. C. (2000) Comparative investigation of electrochemical cholinesterase biosensors for pesticide determination. *Anal. Chim. Acta* **404**(1), 55–65.

Izaoumen, N. Bouchta, D. Zejli, H. Kaoutit, M. E., Stalcup, A. M. and Temsamani, K. R. (2005) Electrosynthesis and analytical performances of functionalized poly (pyrrole/beta-cyclodextrin) films. *Talanta* **66**(1), 111–117.

Ji, X., Zheng, J. Xu, J., Rastogi, V., Cheng, T. C., DeFrank J. J. and Leblanc, R. M. (2005) (CdSe)ZnS quantum dots and organophosphorus hydrolase bioconjugate as biosensors for detection of paraoxon. *J. Phys. Chem. B* **10**(9), 3793–3799.

Jimenez-Cadena, G., Riu, J. and Rius, F. X. (2007) Gas sensors based on nanostructured materials. *Analyst.* **132**(11), 1083–1099.

Joseph, R. L. (2006) *Principles of Fluorescence Spectroscopy*, Springer, 3rd Ed, New York, NY, USA.

Joshi, K. A., Tang, J., Haddon, R., Wang, J., Chen, W. and Mulchandani, A. (2005) A disposable biosensor for organophosphorus nerve agents based on carbon nanotubes modified thick film strip electrode. *Electroanal.* **17**(1), 54–58.

Jun, Y., Choi, J.-S. and Cheon, J. (2006) Shape control of semiconductor and metal oxide nanocrustals through nonhydrolytic colloidal routes. *Angew. Chem. Int. Ed.* **45**(21), 3411–3439.

Kamyabi, M. A. and Aghajanloo, F. (2008) Electrocatalytic oxidation and determination of nitrite on carbon paste electrode modified with oxovanadium(IV)-4-methyl salophen. *J. Electroanal. Chem.* **614**(1–2), 157–165.

Kan, J., Pan, X. and Chen, C. (2004) Polyaniline-uricase biosensor prepared with template process. *Biosens. Bioelectron.* **19**(12), 1635–1640.

Kerman, K., Saito, M, Tamiya, E. Yamamura, S. and Takamura, Y. (2008) Nanomaterial-based electrochemical biosensors for medical application, *Trends Anal. Chem.* **27**(7), 585–592.

Khan, R. and Dhayal, M. (2008) Nanocrystalline bioactive TiO_2–chitosan impedimetric immunosensor for ochratoxin-A. *Comm.* **10**(3), 492–495.

Kim, A., Li, C. Jin, C., Lee, K. W., Lee, S., Shon, K., Park, N., Kim, D., Kang, S., Shim, Y. and Park, J. (2007) A Sensitive and reliable quantification method for Bisphenol A based on modified competitive ELISA method. *Chemosphere* **68**(7), 1204–1209.

Kim, G. Y., Shim, J., Kang, M. S. and Moon, S. H. (2008) Optimized coverage of gold nanoparticles at tyrosinase electrode for measurement of a pesticide in various water sample. *J. Hazard. Mater.* **156**(1–3), 141–147.

Kurosawa, S., Aizawa, H. and Park, J.-W. (2005) Quartz crystal microbalance immunosensor for highly sensitive 2,3,7,8-tetrachlorodibenzo-p-dioxin detection in fly ash from municipal solid waste incinerators. *Analyst.* **130**(11), 1495–1501.

Kuwahara, T., Oshima, K., Shimomura, M. and Miyauchi, S. (2005) Immobilization of glucose oxidase and electron-mediating groups on the film of 3-methylthiophene/thiophene-3-acetic acid copolymer and its application to reagentless sensing of glucose. *Polymer* **46**(19), 8091–8097.

Lange, U., Roznyatovskaya, N. V. and Mirsky, V. M. (2008) Conducting polymers in chemical sensors and arrays. *Anal. Chim. Acta.* **614**(1), 1–26.

Law, K. A. and Higson, J. (2005) Sonochemically fabricated acetylcholinesterase micro-electrode arrays within a flow injection analyser for the determination of organophosphate pesticides. *Biosens. Bioelectron.* **20**(10), 1914–1924.

Lei, Y., Mulchandani, P., Chen, W. and Mulchandani, A. (2005) Direct determination of p-nitrophenyl substituent organophosphorus nerve agents using a recombinant Pseudomonas putida JS444-modified Clark oxygen electrode. *J. Agric. Food Chem.* **53**(3), 524–527.

Lei, Y., Mulchandani, P., Chen, W., Mulchandani, A. (2007) Biosensor for direct determination of fenitrothion and EPN using recombinant Pseudomonas putida JS444 with surface-expressed organophosphorous hydrolase. 2. Modified carbon paste electrode. *Appl. Biochem. Biotechnol.* **136**(3), 243–50.

Leung, M. K. P., Chow, C. F. and Lam, M. H. W. (2001) A sol-gel derived molecular imprinted luminescent PET sensing material for 2,4-dichlorophenoxyacetic acid. *J. Mater. Chem.* **11**(12), 2985–2991.

Li, C., Wang, C., Guan, B., Zhang, Y. and Hu, S. (2005) Electrochemical sensor for the determination of parathion based on *p-tert*-Butylcalix[6]-arene-1,4-crown-4 sol-gel film and its characterization by electrochemical methods. *Sens. Actuators B* **107**(1), 411–417.

Li, Y. F., Liu, Z. M., Liu, Y. Y., Yang, Y. H., Shen, G. L. and Yu, R. Q. (2006) A mediator-free phenol biosensor based on immobilizing tyrosinase to ZnO nanoparticles. *Anal. Biochem.* **349**(1), 33–40.

Liao, Y., Wang, W. and Wang, B. (1999) Building fluorescent sensors by template polymerization: The preparation of a fluorescent sensor for l-tryptophan. *Bioorg. Chem.* **27**(6), 463–476.

Lin, H. Y., Hsu, C. Y. Thomas, J. L. Wang, S. E., Chen, H. C. and Chou, T. C. (2006) The microcontact imprinting of proteins: The effect of cross-linking monomers for lysozyme, ribonuclease A and myoglobin. *Biosens. Bioelectron.* **22**(4), 534–543.

Lin, J. M. and Yamada, M. (2001) Chemiluminescent flow-through sensor for 1,10-phenanthroline based on the combination of molecular imprinting and chemiluminescence. *Analyst* **126**(6), 810–815.

Liu, G. and Lin, Y. (2007) Nanomaterial labels in electrochemical immunosensors and immunoassays. *Talanta* **74**(3), 308–317.

Liu, N., Cai, X., Lei, Y., Zhang, Q., Chan-Park, M. B., Li, C., Chen, W. and Mulchandani, A. (2007) Single-walled carbon nanotube based real-time organophosphate detector. *Electroanalysis* **19**(5), 616–619.

Liu, K., Wei, W., Zeng, J. X., Liu, X. Y. and Gao, Y. P. (2006) Application of a novel electrosynthesized polydopamine-imprinted film to the capacitive sensing of nicotine. *Anal. Bioanal. Chem.* **38**(4), 724–729.

Liu, M., Hashi, Y., Pan, F., Yao, J. , Song, G. and Lin, J. (2006a) Automated on-line liquid chromatography-photodiode array-mass spectrometr method with dilution line for the determination of bisphenol A and 4-octylphenol in serum . *J. Chromatogr. A* **1133**(1–2), 142–148.

Liu, Z., Liu, Y., Yang, H., Yang, Y., Shen, G. and Yu, R. (2005a) A phenol biosensor based on immobilizing tyrosinase to modified core-shell magnetic nanoparticles supported at a carbon paste electrode. *Anal. Chim. Acta* **533**(1), 3–9.

Liu, J. F., Wang, X., Peng, Q. and Li, Y. D. (2005b) Vanadium pentoxide nanobelts: Highly selective and stable ethanol sensor materials. *Adv. Mater.* **17**(6), 764–767.

Lopéz, M. S.-P., Lopéz-Cabarcos, E. and Lopéz-Ruiz, B. (2006) Organic phase enzyme electrodes. *Biomol. Eng.* **23**(4), 135–147.

Lu, P., Teranishi, T., Asakura, K., Miyake, M. and Toshima, N. (1999) Polymer-protected Ni/Pd bimetallic nano-clusters: Preparation, characterization and catalysis for hydrogenation of nitrobenzene. *J. Phys. Chem. B* **10**(44), 9673–9682.

Marchesini, G. R., Koopal, K., Meulenberg, E., Haasnoot, W. and Irth. H (2007) Spreeta-based biosensor assays for endocrine disruptors, *Biosen. Bioelectron.* **22**(9–10), 1908–1915.

Marchesini, G. R., Meulenberg, E., Haasnoot, W. and Irth, H. (2005) Biosensor immunoassays for the detection of bisphenol A. *Anal. Chim. Acta* **528**(1), 37–45.

Marchesini, G. R., Meulenberg, E., Haasnoot, W., Mizuguchi, M. and Irth, H. (2006) Biosensor recognition of thyroid-disrupting chemicals using transport proteins. *Anal. Chem.* **78**(4), 1107–1114.

Matsui, J., Akamatsu, K., Nishiguchi, S., Miyoshi, D., Nawafune, H., Tamaki, K. and Sugimoto, N. (2004) Composite of Au nanoparticles and molecularly imprinted polymer as a sensing material. *Anal. Chem.* **76**(5), 1310–1315.

Matsui, J., Higashi, M. and Takeuchi, T. (2000) Molecularly imprinted polymer as 9-ethyladenine receptor having a porphyrin-based recognition center. *J. Am. Chem. Soc.* **12**(21), 5218–5219.

Mauriz, E., Calle, A., Lechuga, L. M., Quintana, J., Montoya, A. and Manclus, J. J. (2006a) Real-time detection of chlorpyrifos at part per trillion levels in ground, surface and drinking water samples by a portable surface plasmon resonance immunosensor. *Anal. Chim. Acta* **561**(1–2), 40–47.

Mauriz, E. Calle, A. Manclus, J. J., Montoya, A., Escuela, A. M., Sendra, J. R. and Lechuga, L. M. (2006b) Single and multi-analyte surface plasmon resonance assays for simultaneous detection of cholinesterase inhibiting pesticides. *Sens. Actuators B* **118**(1–2), 399-407.

Melikhova, E. V., Kalmykova, E. N., Eremin, S. A. and Ermolaeva, T. N. (2006) Using a piezoelectric flow immunosensor for determining sulfamethoxazole in environmental samples. *J. Anal. Chem.* **61**(7), 687-693.

Migdalski, J., Blaz, T., Paczosa, B. and Lewenstam, A. (2003) Magnesium and calcium-dependent membrane potential of poly(pyrrole) films doped with adenosine triphosphate. *Microchim. Acta* **143**(2–3), 177-185.

Milson, E. V., Novak, J., Oyama, M. and Marken, F. (2007) Electrocatalytic oxidation of nitric oxide at TiO2-Au nanocomposite film electrodes. *Electrochem. Commun.* **9**(3), 436–442.

Mita, D. G., Attanasio, A., Arduini, F., Diano, N., Grano, V., Bencivenga, U., Rossi, S., Amine, A. and Moscone, D. (2007) Enzymatic determination of BPA by means of tyrosinase immobilized on different carbon carriers. *Biosens. Bioelectron.* **23**(1), 60–65.

Mottaleb, M. A., Usenko, S., O'Donnell, J. G., Ramirez, A. J., Brooks, B. W. and Chambliss, C. K. (2009) Gas chromatography–mass spectrometry screening methods for select UV filters, synthetic musks, alkylphenols, an antimicrobial agent, and an insect repellent in fish. *J. Chromatogr. A* **1216**(5), 815–823.

Mousavi, Z., Alaviuhkola, T., Bobacka, J., Latonen, R. M., Pursiainen, J. and Ivaska, A. (2008) Electrochemical characterization of poly (3,4-ethylenedioxythiophene) (PEDOT) doped with sulfonated thiophenes. *Electrochim. Acta* **53**(11), 3755-3762.

Mulchandani, A., Mulchandani, P., Chauhan, S., Kaneva, I. and Chen, W. (1998a) A potentiometric microbial biosensor for direct determination of organophosphate nerve agents. *Electroanalysis* **10**(11), 733-737.

Mulchandani, A., Mulchandani, P., Kaneva, I. and Chen, W. (1998b) Biosensor for direct determination of organophosphate nerve agents using recombinant Escherichia coli with surface-expressed organophosphorus hydrolase. 1. Potentiometric microbial electrode. *Anal. Chem.* **70**(19), 4140-4145.

Nakao, Y. and Kaeriyama, K. (1989) Adsorption of surfactant-stabilized colloidal noble metals by ion-exchange resins and their catalytic activity for hydrogenation. *J. Colloids Interf. Sci.* **131**(1), 186–191.

Nakata, H. Kannan, K., Jones, P. D. and Giesy, J. P. (2005) Determination of fluoroquinolone antibiotics in wastewater effluents by liquid chromatography–mass spectrometry and fluorescence detection, *Chemosphere* **58**(6), 759–766.

Navarro-Villoslada, F., Urraca, J. L., Moreno-Bondi, M. C. and Orellana, G. (2007) Zearalenone sensing with molecularly imprinted polymers and tailored fluorescent probes. *Sens. Actuators B: Chem.* **12**(1), 67-73.

Nistor, C., Rose, A., Farré, M., Stoica, L., Wollenberger, U., Ruzgas, T., Pfeiffer, D., Barcelo, D., Gorton, L. and Emneus, J. (2002) In-field monitoring of cleaning efficiency in waste water treatment plants using two phenol-sensitive biosensors. *Anal. Chim. Acta* **456**(1), 3–17.

Park, J., Kurosawa, S., Aizawa, H., Goda, Y., Takai, M. and Ishihara, K. (2006) Piezoelectric immunosensor for bisphenol A based on signal enhancing step with 2-methacrolyloxyethyl phosphorylcholine polymeric nanoparticles. *Analyst* **31**(1), 155–162.

Petrovic, M., Hernando, M. D., Diaz-Cruz, M. S. and Barcelo, D. (2005) Liquid chromatography–tandem mass spectrometry for the analysis of pharmaceutical residues in environmental samples: A review, *J. Chromatoga. A* **1067**(1–2), 1–14.

Pillay, J. and Ozoemena, K. I. (2007) Efficient electron transport across nickel powder modified basal plane pyrolytic graphite electrode: Sensitive detection of sulfohydryl degradation products of the V-type nerve agents. *Electrochemistry* **9**(7), 1816–1823.

Pita, M. T. P., Reviejo, A. J., Manuel de Villena, F. J., Pingarron, J. M. (1997) Amperometric selective biosensing of dimethyl- and diethyldithiocarbamates based on inhibition processes in a medium of reversed micelles. *Anal. Chim. Acta* **340**(1–3), 89–97.

Portnov, S., Aschennok, A., Gubskii, A., Gorin, D., Neveshkin, A., Klimov, B., Nefedov, A. and Lomova, M. (2006) An automated setup for production of nanodimensional coatings by the polyelectrolyte self-assembly method. *Instrum. Exp. Technol.* **4**(6), 849–854.

Potyrailo, R. A. and Mirsky, V. M. (2008) Combinatorial and high-throughput development of sensing materials: The first 10 years. *Chem. Rev.* **108**(2), 770–813.

Prakash, R. Srivastava, R. and Pandey, P. (2002) Copper(II) ion sensor based on electro-polymerized undoped conducting polymers. *J. Solid State Electrochem.* **6**(3), 203–208.

Prasad, K., Prathish, K. P., Gladis, J. M., Naidu, G. R. K. and Prasada Rao, T. (2007) Molecularly imprinted polymer (biomimetic) potentiometric sensor for atrazine. *Sens. Actuators B: Chem.* **123**(1), 65–70.

Pringsheim, E. Terpetschnig, E. Piletsky, S. A. and Wolfbeis, O. S. (1999) A polyaniline with near-infrared optical response to saccharides. *Adv. Mater.* **11**(10), 865–868.

Pumera, M., Sanchez, S., Ichinose, I. and Tang, J. (2007) Electrochemical nanobio-sensors. *Sens. Actuators, B* **123**(2), 1195–1205.

Quist, A. P., Pavlovic, E. and Oscarsson, S. (2005) Recent advances in microcontact printing. *Anal. Bioanal. Chem.* **381**(3), 591–600.

Rajasekar, S., Rajasekar, R. and Narasimham, K. C. (2000) Acetobacter peroxydans based electrochemical biosensor for hydrogen peroxide. *Bull. Electrochem.* **16**(1), 25–28.

Rahman, M. A., Kwon, M.-S., Won, Choe E. S. and Shim, Y.-B. (2005) Functionalized conducting polymer as an enzyme-immobilizing substrate: an amperometric glutamate microbiosensor for in vivo measurements. *Anal. Chem.* **77**(15), 4854–4860.

Rahman, M. A. R., Shiddiky, M. J. A., Park, J.-S. and Shim, Y.-B. (2007) An impedimetric immunosensor for the label-free detection of bisphenol A. *Biosen. Bioelectron.* **22**(11), 2464–2470.

Rico, M. A. G., Olivares-Marin, M. and Gil, E. P. (2009) Modification of carbon screen-printed electrodes by adsorption of chemically synthesized Bi nanoparticles for the voltammetric stripping detection of Zn(II), Cd(II) and Pb(II), *Talanta* **80**(9), 631–635.

Rivas, G. A., Rubianes, M. D., Rodriguez, M. C., Ferreyra, N. E., Luque, G. L., Pedano, M. L., Miscoria, S. A. and Parrado, C. (2007) Carbon nanotubes for electrochemical biosensing. *Talanta* **74**(3), 291–307

Rodrigues, A. M., Ferreira, V., Cardoso, V. V., Ferreira, E. and Benoliel, M. J. (2007) Determination of several pesticides in water by solid-phase extraction, liquid chromatography and electrospray tandem mass spectrometry. *J. Chromatogr. A* **1150**(1–2), 267–278.

Rodriguez-Mozaz, S., Lopéz de Alda, M. J. and Barcelo, D. (2006a) Biosensors as useful tools for environmental analysis and monitoring, *Anal. Bioanal. Chem.* **386**(4), 1025–1041.

Rodriguez-Mozaz, S., Lopéz de Alda, M. J. and Barcelo, D. (2006b) Fast and simultaneous monitoring of organic pollutants in a drinking water treatment plant by a multi-analyte biosensor followed by LC-MS validation. *Talanta* **69**(2), 377–384.

Rodriguez-Monaz, S. Lopéz de Alda, M. and Barcelo, D. (2005) Analysis of bisphenol A in natural waters by means of an optical immunosensor. *Water Res.* **39**(20), 5071–5079.

Sacks, V., Eshkenazi, I., Neufeld, T., Dosoretz, C. and Rishpon, J. (2000) Immobilized parathion hydrolase: An amperometric sensor for parathion. *Anal. Chem.* **72**(9), 2055–2058.

Salimi, A., Hallaj, R., Soltanian, S. and Mamkhezri, H. (2007) Nanomolar detection of hydrogen peroxide on glassy carbon electrode modified with electrodeposited cobalt oxide nanoparticles. *Anal. Chim. Acta* **594**(1), 24–31.

Sanchez-Acevedo Z. C., Riu, J. and Rius, F. X. (2009) Fast picomolar selective detection of bisphenol A in water using a carbon nanotube field effect transistor functionalized with estrogen receptor-α. *Biosen. Bioelectron.* **24**(9), 2842–2846.

Saraji, M. and Esteki, N. (2008) Analysis of carbamate pesticides in water samples using single-drop microextraction and gas chromatography-mass spectrometry. *Anal. Bioanal. Chem.* **391**(3), 1091–1100.

Sellergren, B. (2001) *Molecularly Imprinted Polymers: Man-made Mimics of Antibodies and their Application in Analytical Chemistry*. Elsevier, Amsterdam, Netherlands.

Seol, H., Shin, S. C. and Shim, Y.-B. (2004) Trace analysis of Al(III) ions based on the redox current of a conducting polymer. *Electroanalysis* **16**(24), 2051–2057.

Sha, Y., Qian, L., Ma, Y., Bai, H. and Yang, X. (2006) Multilayer films of carbon nanotubes and redox polymer on screen-printed carbon electrodes for electrocatalysis of ascorbic acid. *Talanta* **70**(3), 556–560.

Sharma, S. K., Singhal, R., Malhotra, B. D., Sehgal, N. and Kumar, A. (2004) Langmuir-Blodgett film based biosensor for estimation of galactose in milk. *Electrochim. Acta* **49**(15), 2479–2485.

Shelke, C. R., Kawtikwar, P. K., Sakarkar, D. M. and Kulkarni, N. P. (2008) Synthesis and characterization of MIPs – a viable commercial venture. *Pharmaceutical Reviews* **6**(5) (http://www.pharmainfo.net/reviews/synthesis-and-characterization-mips-viable-commercial-venture) (Accessed on Jan, 2010).

Shiddiky, M. J. A., Rahman, M., Cheol, C. S. and Shim, Y. B. (2008) Fabrication of disposable sensors for biomolecule detection using hydrazine electrocatalyst. *Anal. Biochem.* **379**(2), 170.

Shughart, E. L., Ahsan, K., Detty, M. R. and Bright, F. V. (2006) Site selectively templated and tagged xerogels for chemical sensors. *Anal. Chem.* **78**(9), 3165–3170.

Singh, S., Chaubey, A. and Malhotra, B. D. (2004) Amperometric cholesterol biosensor based on immobilized cholesterol esterase and cholesterol oxidase on conducting polypyrrole films. *Anal. Chim. Acta* **502**(2), 229–234.

Singh, S., Solanki, P. R., Pandey, M. K. and Malhotra, B. D. (2006) Covalent immobilization of cholesterol esterase and cholesterol oxidase on polyaniline films for application to cholesterol biosensor. *Anal. Chim. Acta* **568**(1–2), 126–132.

Skretas, G. and Wood, D. W. (2005) A bacterial biosensor of endocrine modulators. *J. Mol. Biol.* **349**(3), 464–474.

Somers, R. C., Bawendi, M. G. and Nocera, D, G. (2007) CdSe nanocrystal based chem-bio-sensors. *Chem. Soc. Rev.* **36**(4), 579–591.

Sreenivasan, K. (2007) Synthesis and evaluation of multiply templated molecularly imprinted polyaniline. *J. Mater. Sci.* **42**(17), 7575-7578.

Suri, C. R., Boro, R., Nangia, Y., Gandhi, S., Sharma, P., Wangoo, N., Rajesh, K. and Shekhawat, G. S. (2009) Immunoanalytical techniques for analyzing pesticides in the environment, *Trends Anal. Chem.* **28**(1), 29–39.

Székács, A., Trummer, N., Adányi, N., Váradi, M. and Szendro, I. (2003) Development of a non-labeled immunosensor for the herbicide trifluralin *via* optical waveguide light mode spectroscopic detection. *Anal. Chim. Acta* **487**(1), 31–42.

Taranova, L. A., Fesay, A. P., Ivashchenko, G. V., Reshetilov, A. N., Winther-Nielsen, M. and Emneus, J. (2004) Comamonas testosteroni strain TI as a potential base for a microbial sensor detecting surfactants. *Appl. Biochem. Microbiol.* **40**(4), 404–408.

Thompson, J. M. and Bezbaruah, A. N. (2008) *Selected Pesticide Remediation with Iron Nanoparticles: Modeling and Barrier Applications.* Technical Report No. ND08-04. North Dakota Water Resources Research Institute, Fargo, ND.

Trojanowicz, M. (2006) Analytical applications of carbon nanotubes: a review. *Trends Anal. Chem.* **25**(5), 480–489.

Tudorache, M. and Bala, C. (2007) Biosensors based on screen-printing technology, and their applications in environmental and food analysis. *Anal. Bioanal. Chem.* **388**(3), 565–578.

USEPA (2010a) What are endocrine disruptors? http://www.epa.gov/endo/pubs/edspoverview/whatare.htm Accessed January 2010.

USEPA (2010b) Drinking water contaminants http://www.epa.gov/safewater/contaminants/index.html Accessed January 2010.

Vandevelde, F., Leichle, T., Ayela, C., Bergaud, C., Nicu, L. and Haupt, K. (2007) Direct patterning of molecularly-imprinted microdot arrays for sensors and biochips. *Langmuir* **23**(12), 6490–6493.

Vaseashta, A., Vaclavikova, M., Vaseashta, S., Gallios, G., Roy, P. and Pummakarnchana, O. (2007) Nanostructures in environmental pollution detection, monitoring, and remediation. *Sci Tech. Adv. Mat.* **8**(1–2), 47–59.

Vazquez, M., Bobacka, J., and Ivaska, A. (2005) Potentiometric sensors for Ag+ based on poly(3-octylthiophene) (POT). *J. Solid State Electrochem.* **9**(12), 865–873.

Vedrine, C., Fabiano, S. and Tran-Minh, C. (2003) Amperometric tyrosinase based biosensor using an electrogenerated polythiophene film as an entrapment support. *Talanta* **59**(3), 535–544.

Vieno, N. M., Tuhkanen, T. and Kronberg, L. (2006) Analysis of neutral and basic pharmaceuticals in sewage treatment plants and in recipient rivers using solid phase extraction and liquid chromatography–tandem mass spectrometry detection, *J. Chromatogr. A* **1134**(1–2), 101–111.

Voicu, R., Faid, K., Farah, A. A., Bensebaa, F., Barjovanu, R., Py, C. and Tao Y. (2007) Nanotemplating for two-dimensional molecular imprinting. *Langmuir* **23**(10), 5452–5458.

Volf, R., Kral, V., Hrdlicka, J., Shishkanova, T. V., Broncova, G., Krondak, M., Grotschelova, S., St'astny, M., Kroulik, J., Valik, M., Matejka, P. and Volka, K., (2002) Preparation, characterization and analytical application of electropolymerized films. *Solid State Ionics* **154**(Part B Sp. Iss. SI), 57–63.

Wallace, G. G., Spinks, G. M., Kane-Maguire, L. A. P. and Teasdale, P. R. (2003) *Conductive Electroactive Polymers: Intelligent Materials Systems.* CRC Press, Boca Raton, FL, USA.

Wang, J. and Lin, Y. (2008) Functionalized carbon nanotubes and nanofibres for biosensing application, *Trends Anal. Chem.* **27**(7), 619–626.

Wang, J., Musameh, M. and Laocharoensuk, R. (2005) Magnetic catalytic nickel particles for on-demand control of electrocatalytic processes. *Electrochem. Commun.* **7**(7), 652–656.

Wang, J., Scampicchio, M., Laocharoensuk, R., Valentini, F., Gonzalez-Garcia, O. and Burdick, J. (2006) Magnetic tuning of the electrochemical reactivity through controlled surface orientation of catalytic nanowires. *J. Am. Chem. Soc.* **128**(4), 4562–4563.

Wang, S., Wu, Z., Zhang, F. Q. S., Shen, G.and Yu, R. (2008) A novel electrochemical immunosensor based on ordered Au nano-prickle clusters, *Biosens. Bioelectron.* **24**(4), 1026–1032.

Wen, Y., Zhou, B., Xu, Y., Jin, S. and Feng, Y. (2006) Analysis of estrogens in environmental waters using polymer monolith in-polyether ether ketone tube solid-phase microextraction combined with high-performance liquid chromatography. *J. Chromatogr. A* **1133**(1–2), 21–28.

Whitcombe, M. J. and Vulfson, E. N. (2001) Imprinting polymer. *Adv. Mater.* **13**(7), 467–478.

Whitcombe, M. J., Rodriguez, M. E., Villar, P. and Vulfson, E. N. (1995) A new method for the introduction of recognition site functionality into polymers prepared by molecular imprinting: Synthesis and characterization of polymeric receptors for cholesterol. *J. Am. Chem. Soc.* **117**(27), 7105–7111.

Widstrand, C., Yilmaz, E., Boyd, B., Billing, J. and Rees, A. (2006) Molecularly imprinted polymers: A new generation of affinity matrices. *American Lab.* **38**(19), 12–14.

Wilmer, M., Trau, D., Renneberg, R. and Spener, F. (1997) Amperometric immunosensor for the detection of 2,4-dichlorophenoxyacetic acid (2,4-D) in water. *Anal. Lett.* **30**(3), 515–525.

Wulff, G., Vesper, W., Grobe-Einsler, R. and Sarhan, A. (1977) Enzyme-analogue built polymers, 4. On the synthesis of polymers containing chiral cavities and their use for the resolution of racemates. *Makromol. Chem.* **178**(10), 2799– 2816.

Yan, M. and Kapua, A. (2001) Fabrication of molecularly imprinted polymer microstructures. *Anal. Chim. Acta* **435**(1), 163–167.

Yan, M. and Ramstrom, O. (2005) *Molecularly Imprinted Materials: Science and Technology*. Marcel Dekker, New York, NY.

Yang, L., Wanzhi, W., Xia, J., Tao, H. and Yang, P. (2005) Capacitive biosensor for glutathione detection based on electropolymerized molecularly imprinted polymer and kinetic investigation of the recognition process. *Electroanalysis* **17**(11) 969–977.

Yao, S., Xu, J., Wang, Y., Chen, X., Xu, Y. and Hu, S. (2006) A highly sensitive hydrogen peroxide amperometric sensor based on MnO2 nanoparticles and dihexadecyl hydrogen phosphate composite film. *Anal. Chim. Acta* **557**(1–2), 78–84.

Yin, H. S., Zhou, Y. and Ai, S.-Y. (2009) Preparation and characteristic of cobalt phthalocyanine modified carbon paste electrode for bisphenol A detection. *J. Electroanalytical Chem.* **626**(1–2), 80–88.

Yu, J. C. C., Krushkova, S., Lai, E. P. C. and Dabek-Zlotorzynska, E. (2005) Molecularly-imprinted polypyrrole-modified stainless steel frits for selective solid phase preconcentration of ochratoxin A. *Anal. Bioanal. Chem.* **382**(7), 1534–1540.

Yuan, J., Guo, W. and Wang, W. (2008) Utilizing a CdTe quantum dots-enzyme hybrid system for the determination of both phenolic compounds and hydrogen peroxide. *Anal. Chem.* **80**(4), 1141–1145

Zacco, E., Pividori, M. I., Alegret, S., Galve, R. and Marco, M.-P. (2006) Electrochemical magnetoimmunosensing strategy for the detection of pesticides residues. *Anal. Chem.* **78**(6), 1780–1788.

Zanganeh, A. R. and Amini, M. K. (2007) A potentiometric and voltammetric sensor based on polypyrrole film with electrochemically induced recognition sites for detection of silver ion. *Electrochim. Acta* **52**(11), 3822–3830.

Zhang, W. X. (2003) Nanoscale iron particles for environmental remediation: An overview. *J. Nanopart. Res.* **5**(3–4), 323–332.

Zhang, J., Wang, H., Liu, W., Bai, L., Ma, N. and Lu, L. (2008) Synthesis of molecularly imprinted polymer for sensitive penicillin determination in milk. *Anal. Lett.* **41**(18), 3411–3419.

Zhang, L., Zhou, Q., Liu, Z., Hou, X., Li, Y. and Lv, Y. (2009) Novel Mn_3O_4 Micro-octahedra: Promising cataluminescence sensing material for acetone, *Chem. Mater.* **21**(21), 5066–5071.

Zhang, Z., Haiping, L., Li, H., Nie, L. and Yao, S. (2005) Stereoselective histidine sensor based on molecularly imprinted sol-gel films. *Anal. Biochem.* **336**(1), 108–116.

4

Nanofiltration: Membrane Materials, Separation and Fouling

C. Y. Chang

4.1 INTRODUCTION OF NANOFILTRATION

Membrane separation is addressed as a pressure driven process. Pressure driven processes are commonly divided into four overlapping categories of increasing selectivity: microfiltration (MF), ultrafiltration (UF), nanofiltration (NF) and hyperfiltration or reverse osmosis (RO). MF can be used to remove bacteria and suspended solids with pore sizes of 0.1 to micron. UF will remove colloids, viruses and certain proteins with pore size of 0.0003 to 0.1 microns. NF relies on physical rejection based on molecular size and charge. Pore sizes are in the range of 0.001 to 0.003 microns. RO has a pore size of about 0.0005 microns and can be used for desalination. A membrane filtration spectrum is shown in Figure 4.1.

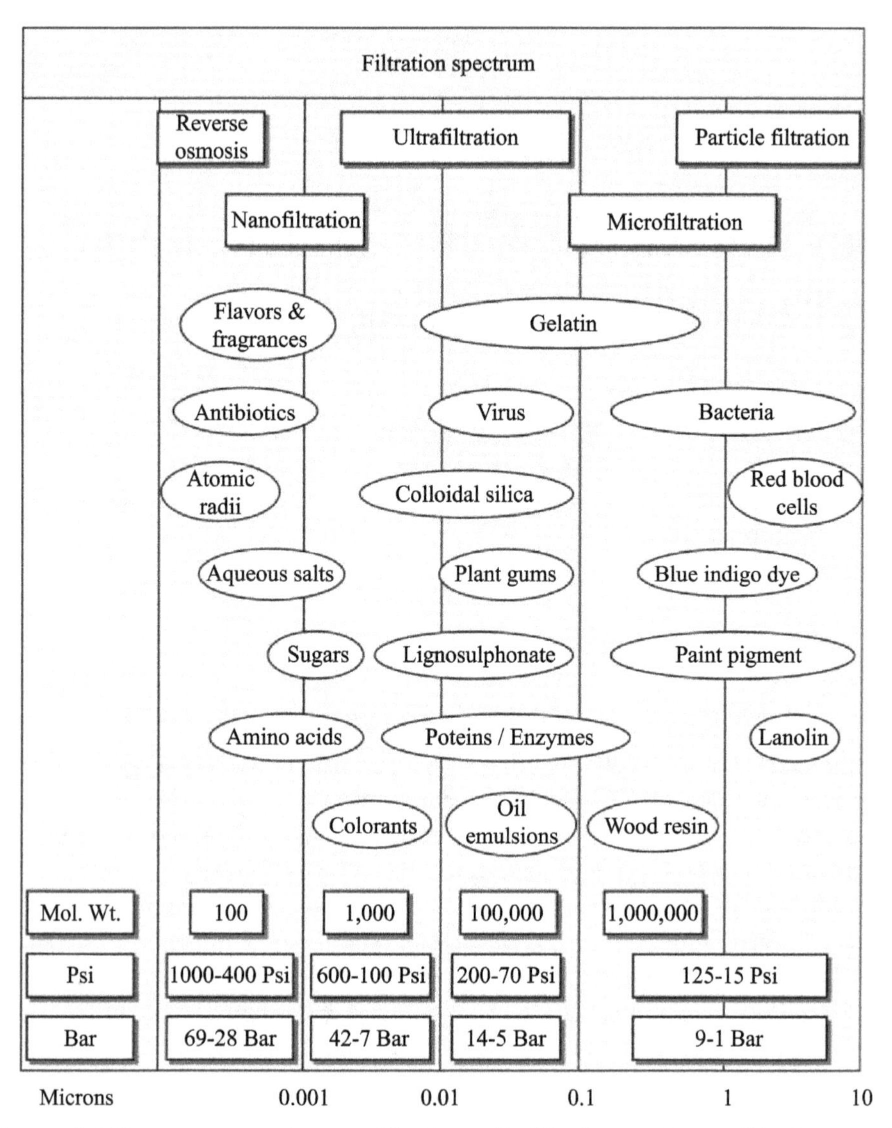

Figure 4.1 Reverse osmosis, nanofiltration, ultrafiltration and microfiltration are all related processes differing principally in the average pore diameter of the membrane and driving pressure

High pressures are required to cause water to pass across the membrane from a concentrated to dilute solution. In general, driving pressure increases as selectivity increases. Clearly it is desirable to achieve the required degree of separation (rejection) at the maximum specific flux (membrane flux/driving pressure). Separation is accomplished by MF membranes and UF membranes via mechanical sieving, while capillary flow or solution diffusion is responsible for separation in NF membranes and RO membranes.

Driven by more stringent water and wastewater treatment standards, applications employing membrane processes are increasing rapidly. Nanofiltration (NF), in particular, has been increasingly considered as an ecologically suited, reliable and affordable technique for the production of high quality water from unconventional sources such as brackish water, polluted surface water, and secondary treated effluent where micro-pollutants are to be removed (Dueom and Cabassud, 1999; Nghiem *et al.,* 2004; Van der Bruggen *et al.,* 2008).

The growth of nanofiltration study and application can be explained by a combination of (1) growing demand of water quantity and quality, (2) better manufacturing technology of the membranes, (3) lower prices of membranes due to wide variety of applications, and (4) more stringent standards, e.g., in the drinking water industry.

The history of nano-filtration (NF) dates back to the 1970s when reverse osmosis (RO) membranes with a reasonable water flux operating at relatively low pressures were developed. Hence, the high pressures traditionally employed in reverse osmosis resulted in a considerable energy cost. However, the permeate quality of RO was very good, and often even beyond expectations. Therefore, membranes with lower rejections of dissolved components, but with higher water permeability would be a great improvement for separation technology. Such ''low-pressure reverse osmosis membranes'' became known as nanofiltration membranes.

By the second half of the 1980s, nanofiltration had become established, and the first applications were reported (Conlon and McClellan, 1989; Eriksson, 1988). Nowadays NF membrane is often referred to as ''softening'' membrane because it is very good at rejecting hardness while letting smaller ions like sodium and chloride pass (Duran and Dunkelberger, 1995; Fu *et al.,* 1994). Since then, the application range of nanofiltration has extended tremendously. New possibilities were discovered for drinking water as well as wastewater treatment and process water production, providing answers to new challenges such as arsenic removal (Waypa, 1997; Brandhuber and Amy, 1998; Urase *et al.,* 1998; Košutić *et al.,* 2005; Shih, 2005; Xia *et al.,* 2007), removal of pesticides, endocrine disruptors and chemicals (Nghiem *et al.,* 2004; Causserand *et al.,* 2005; Jung *et al.,* 2005; Košutić *et al.,* 2005; Xu *et al.,* 2005; Zhang *et al.,* 2006; Yoon *et al.,* 2007), and partial desalination (Al-Sofi *et al.,* 1998; Hassan *et al.,* 1998; Hassan *et al.,* 2000; Semiat, 2000).

4.2 NANOFILTRATION MEMBRANE MATERIALS

NF membranes are generally classified into two major groups, organic and ceramic, according to their material properties. Today, organic NF membranes derived from polymeric materials are commercially available. They are applied in various fields such as drinking, process and wastewater treatment.

Two different techniques have been adopted for the development of polymer NF membranes: the phase-inversion method for asymmetric membranes (Jian *et al.,* 1999; Kim *et al.,* 2001), and interfacial polymerization (Rao *et al.,* 1997; Roh *et al.,* 1998; Jegal *et al.,* 2002; Kim *et al.,* 2002; Lu *et al.,* 2002; Song *et al.,* 2005; Verissimo *et al.,* 2005) and other coating techniques for thin film composite (TFC) membranes (Dai *et al.,* 2002; Moon *et al.,* 2004). The TFC membrane approach has some key advantages relative to asymmetric membrane approach because the selective layer and the porous support layer can be optimized separately (Petersen, 1993). Many commercial TFC membranes were prepared on polysulfone (PSf) substrate, because it has excellent chemical resistance and mechanical strength.

In general, hydrophobic polymers, such as polysulfones, polypropylene, and PVDF are widely used as membrane materials due to their good chemical resistance, superior thermal and mechanical properties. However, the affinity of organics for the membranes in a feed solution can readily cause fouling of hydrophobic materials (Ying *et al.,* 2003).

Various means have been employed in order to minimize fouling. Basically, there are either operational procedures applied during the membrane processes or fundamental modifications targeting the membrane material itself. Improvement of membrane productivity by higher flux membranes has been achieved through the development of thin film composite with very thin selective skin layer. Other methods which have been used to improve membrane performance include development of mixed-matrix membrane materials including hybrid organic-inorganic materials and surface modification of membranes. These improvements are not only limited to higher flux at lower operating pressures but also in terms of less fouling propensity, higher chlorine tolerance and increased solvent resistance (Nunes and Peinemann, 2001).

Several methods have been reported with the potential to reduce or eliminate adhesive fouling by changing the membrane surface chemistry. These methods include: (1) physically coating water soluble polymers or charged surfactants onto the membrane surface for temporary surface modification (Kim *et al.,* 1988; Jönsson and Jönsson, 1991), (2) forming ultrathin films on the membrane using Langmuir–Blodgett (LB) techniques (Kim *et al.,* 1989), (3) coating hydrophilic polymers on the membrane using heat curing (Stengaard, 1988;

Hvid *et al.*, 1990), (4) grafting monomers to the membranes by electron beam irradiation (Keszler *et al.,* 1991; Kim *et al.,* 1991), and (5) photografting monomers to the membrane using UV irradiation (Nystrom and Jarvinen, 1991; Yamagishi *et al.,* 1995; Ulbricht *et al.,* 1996).

However, for NF membranes in particular, the role of membrane charge in separations of ions has also been studied considerably. Further understanding of the exact nature of charge formation, as well as how they contribute towards rejection (Tay *et al.,* 2002) will definite leads towards significant improvement in NF membrane performances.

The development of ceramic membrane structure includes a multi-step synthesis procedure, generally based on sol–gel techniques (Burggraaf and Keizer, 1991). Firstly, a macroporous ceramic membrane support is coated with several mesoporous membrane layers in order to correct its surface roughness. Then, the last mesoporous membrane layer is modified with a very thin microporous top-layer, the actual NF layer with a cut-off value below 1000. Silicon, alumina and titanium oxide-based nanoporous membranes are chemically inert and mechanically stable. They have highly uniform and well-defined pore structures (Desai *et al.,* 1999; Martin *et al.,* 2005; Paulose *et al.,* 2008). Al_2O_3, ZrO_2 and TiO_2 are mostly considered as ceramic membrane materials (Luyten *et al.*, 1997).

Soria and Cominotti (1996) mentioned the commercialization of ceramic NF membranes, consisting of macroporous α-Al_2O_3 supports with TiO_2 top-layers, with an MWCO of ca. 1000. Larbot *et al.* (1994), Alami-Younssi *et al.* (1995) and Baticle *et al.* (1997) reported the preparation and characterization of microporous γ-Al_2O_3 membranes.

Van Gestel *et al.* (2002a) prepared and characterized a porous ceramic multilayer nanofiltration (NF) membrane from high-quality macroporous supports of α- Al_2O_3 (Figure 4.2). Three types of colloidal sol–gel derived mesoporous interlayers including Al_2O_3, TiO_2 and mixed Al_2O_3–TiO_2 were used. Corrosion measurements showed that application of a multilayer configuration including weakly crystallized α-Al_2O_3 layers is restricted to mild aqueous media (pH 3–11) or non-aqueous media (organic solvents). Optimized α-Al_2O_3/γ-Al_2O_3/anatase and α-Al_2O_3/anatase/anatase multilayer configurations show high retentions for relatively small organic molecules (molecular weight cut-off <200).

Zeolite membranes may offer an alternative choice for produced water treatment. Zeolites are crystalline aluminosilicate materials with uniform subnanometer- or nanometer-scale pores. For example, the MFI-type zeolite has a three-dimensional pore system with straight channels in the *b*-direction (5.4A $\times$ 5.6 A) and sinusoidal channels in the *a*-direction (5.1A $\times$ 5.5 A).

Due to the inert property of aluminosilicate crystal, zeolite membranes have superior thermal and chemical stabilities, hence holding great potential for application in difficult separations such as produced water purification and radioactive wastewater treatment. Research results demonstrated that zeolite membrane could separate several kinds of ions from water, methanolic, and ethanolic electrolyte solutions (Murad *et al.,* 1998; Lin and Murad, 2001a; Murad *et al.,* 2004; Murad and Nitche, 2004). Kumakiri *et al.* (2000) reported using an A-type zeolite membrane (pore size 0.42 nm) in RO separation of water–ethanol mixtures. The hydrophilic A-type zeolite membrane showed 44% rejection of ethanol and a water flux of 0.058 kgm^{-2} h^{-1} under an applied feed pressure of 1.5 MPa.

Figure 4.2 FESEM cross-section (50,000×) of a multilayer membrane (Van Gestel *et al.,* 2002a)

Since their discovery by Iijima (1991), carbon nanotubes have received an increasing scientific interest because of their exceptional physical properties. In the past decade, carbon nanotubes (CNTs) have been proposed for use in numerous applications, including electronics, composite materials, fuel cells, sensors, optical devices, and biomedicine. Their use in NF membranes manufacturing, however, is still nascent, with few applications proposed or investigated so far.

Previous studies in this field have focused on the use of CNTs or CNTs functionalized with inorganic nanoparticles for adsorption of inorganic pollutantts and toxic metals from water (Long and Yang, 2001; Li *et al.,* 2002; Li *et al.,* 2003; Peng *et al.,* 2003; Agnihotri *et al.,* 2005; Lu *et al.,* 2005; Di *et al.,* 2006; Gauden *et al.,* 2006; Yang *et al.,* 2006).

A limited number of studies have explored the use of CNTs for filtration and separation applications. Srivastava *et al.* (2004) developed a cylindrical membrane filter composed of radially aligned multi-walled carbon nanotubes

(MWNTs) that formed a CNT layer several hundreds of micrometers thick. It was shown that the MWNT filter was effective in removing hydrocarbons from petroleum wastes as well as bacteria and viruses. Wang *et al.* (2005) prepared a composite polymeric ultrafiltration membrane, with oxidized MWNTs incorporated into the top layer. The composite ultrafiltration membrane demonstrated high retention of oil/water emulsions.

Carbon nanotube growth methods can be classified based on the number of walls in a given tube. First, both multiwalled nanotubes and single-walled nanotubes have been grown via arc-discharge carried out in an inert gas atmosphere between carbon or catalyst-containing carbon electrodes. Nowadays, carbon nanotubes and related materials are produced via a wide variety of processes. Three different processes are usually employed for the manufacture of carbon nanotube membranes including electric arc discharge, laser ablation and Chemical Vapour Deposition (CVD). Out of these, the CVD method is widely used because of its certain characteristics (Terrones, 2004).

Choi *et al.* (2006) prepared multi-walled carbon nanotubes (MWNTs)/polysulfone (PSf) blend membranes by a phase inversion process using *N*-methyl-2-pyrrolidinone (NMP) as a solvent and water as a coagulant. Because of the hydrophiic MWNTs, the surface of the MWNTs/PSf blend membranes appeared to be more hydrophilic than a just PSf membrane. The PSf membrane with 4.0% of MWNTs showed higher flux and rejection than the PSf membrane without MWNTs. The order of the flux according to the contents of MWNTs of the blend membranes was 1.5% > 1.0% > 2.0% > 0.5% > 0.0% > 4.0%. The order of the pore size with the contents of MWNTs was: 4.0% < 0.0% < 0.5% < 2.0% < 1.0% < 1.5%.

Zhang *et al.* (2006) conducted a study for the removal of sodium dodecylbenzene sulfonate (SDBS) using silica/titania nanorods/nanotubes composite membrane with photocatalytic capability. XRD patterns confirmed that the embedding of amorphous silica into nanophase titania matrix helped to increase the thermal stability of titania and control the size of titania particles. SPM micrograph (Figure 4.3) shows silica/titania particles with rod-shaped homogenously distributed on the support of alumina and the silica/titania nanorod is about 5 nm high (the thickness of silica/titania layer). But most (95%) of the pore volume is located in mesopores of diameters ranging from 1.4 to 10 nm. It is these mesopore structures that allow rapid diffusion of various products during UV illumination and enhance the rate of photocatalytic reaction. The results showed that the removal of SDBS achieved 89% after 100 min by combining the photocatalysis with membrane filtration techniques.

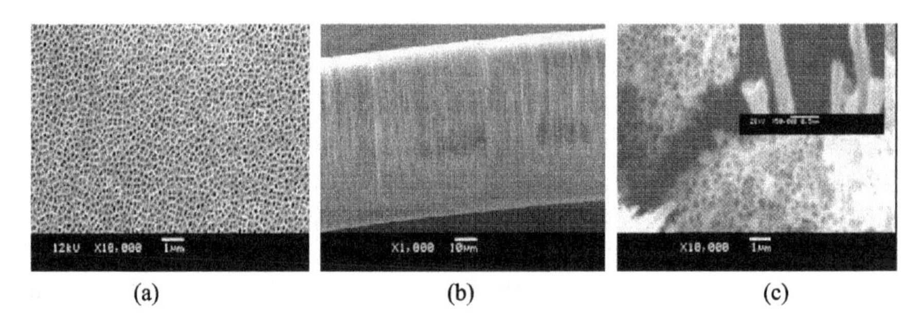

Figure 4.3 SEM photograph of: (a) the surface of 20%-silica/titania composite membrane; (b) the cross-section of 20%-silica/titania composite membrane; (c) 20%-silica/titania nanotubes (×10,000 and ×50,000)

Zhang *et al.* (2008) prepared titanium dioxide (TiO_2) nanotube membrane by grafting anatase TiO_2 nanotubes in the channels of alumina microfiltration (MF) membrane using TiF_4 solution through liquid-phase deposition. The experiment results of continuous filtration under UV irradiation showed that not only HA was rejected and photodegraded by the TiO_2 nanotube membrane, but also the membrane fouling was alleviated dramatically.

Tang *et al.* (2009) prepared chitosan/MWNTs porous membranes and found that the water flux of composite membrane with 10 wt% MWNTs (128.1 L/m^2 h) is 4.6 times that of neat one (27.6 L/m^2 h). In addition, a greatly improved tensile strength of chitosan porous membranes has been achieved by adding MWNTs.

4.3 SEPARATION AND FOULING OF NANOFILTRATION

The extent to which NF membranes are capable of retaining components depends on the NF membrane molecular weight cut-off (MWCO), surface morphology and the nature of the membrane material. Usually, the MWCO is defined as the MW of a solute that was rejected at 90 percent by a specific membrane (Van der Bruggen *et al.,* 1999). However, the value of MWCO generally will be affected by test protocols including solute characteristics, solute concentration, solvent characteristics, as well as flow conditions such as dead-end versus cross-flow filtration. Normally, sieving effects due to steric hindrance will increase for larger molecule and the molecule is rejected by the membrane more often than a smaller molecule. However, the MWCO is only capable of providing a rough estimate of the sieving effect (Mohammad and Ali, 2002; Van der Bruggen and Vandecasteele, 2002). It may also be related to diffusion since a bigger molecule will diffuse more slowly than a smaller molecule.

The desalting degree of a membrane is also frequently used to describe the rejection characteristics of a membrane. Kiso *et al.* (1992; 2000) reported that the membranes with the highest desalting degree showed the highest pesticide rejection. Membrane surface morphology such as porosity and roughness has been regarded as another useful parameter in previous studies to estimate organic compound separation (Košutić *et al.,* 2000; Košutić and Kunst, 2002; Lee *et al.,* 2002). Košutić *et al.* (2000) reported that the membrane porous structure was the dominant parameter in determining the membrane performance, and that solute rejection could be explained by membrane pore size distribution (PSD) and effective number (N) of pores in the upper membrane layer. Scanning electron microscopy (SEM), atomic force microscopy (AFM) and field emission scanning electron microscopy (FESEM) have commonly been used for characterizing membrane surface morphology (Hirose *et al.,* 1996; Chung *et al.,* 2002).

The effects of solute characteristics on the membranes performance have been investigated by many researchers (Kiso *et al.,* 1992; Berg *et al.,* 1997; Van der Bruggen *et al.,* 1998; Kiso *et al.,* 2001a, b; Košutić and Kunst, 2002; Ozaki and Li, 2002; Schutte, 2003). The findings from those studies can be concluded that a quantification of the molecular size (and geometry) of non-charged and non-polar compounds coupled with the pore size of a membrane might be a better descriptor of the rejection than MWCO, MW, or desalting degree since steric hindrance may be an important driving factor in the rejection of molecules by NF membranes (Kiso *et al.,* 2001a; Košutić and Kunst, 2002; Ozaki and Li, 2002; Schutte, 2003). In addition, non-charged compound s with a higher number of methyl groups were reportedly rejected at higher levels than ones with lower numbers of methyl groups (Berg *et al.,* 1997).

Furthermore, several studies confirmed that molecular size parameters such as molecular width, Stokes radii, and molecular mean size have been shown to be a better predictor of steric hindrance effects upon the rejection of solutes by NF membranes than MW (Kiso *et al.,* 1992; Berg *et al.,* 1997; Van der Bruggen *et al.,* 1998; Van der Bruggen *et al.,* 1999; Kiso *et al.,* 2001a; Kiso *et al.,* 2001b; Ozaki and Li, 2002).

Electrostatic interactions between charged solutes and a porous membrane have been frequently reported to be an important rejection mechanism (Wang *et al.,* 1997; Bowen and Mohammad, 1998; Xu and Lebrun, 1999; Childress and Elimelech, 2000; Bowen *et al.,* 2002; Mohammad and Ali, 2002; Wang *et al.,* 2002).

For most thin-film composite (TFC) membranes, in order to minimize the adsorption of negatively charged foulants present in membrane feed waters and increase the rejection of dissolved salts, the membrane skin functionally carries a negative charge (Xu and Lebrun, 1999; Deshmukh and Childress, 2001; Shim *et al.,* 2002). Zeta potentials for most membranes have been observed in many studies to become increasingly more negative as pH is increased and functional groups (such as sulfonic and/or carboxylic acid groups which are deprotonated at neutral pH) deprotonate (Braghetta *et al.,* 1997; Hagmeyer and Gimbel, 1998; Deshmukh and Childress, 2001; Ariza *et al.,* 2002; Lee *et al.,* 2002; Tanninen *et al.,* 2002; Yoon *et al.,* 2002).

Charged organics and dissolved ion rejections by TFC NF membranes are heavily dependent upon the membrane surface charge and therefore feed water chemistry (Berg *et al.,* 1997; Wang *et al.,* 1997; Hagmeyer and Gimbel, 1998; Yoon *et al.,* 1998; Xu and Lebrun, 1999; Childress and Elimelech, 2000; Ozaki and Li, 2002; Wang *et al.,* 2002). Increasing the pH increased the negative surface charge of the membrane as confirmed by others (Braghetta *et al.,* 1997; Deshmukh and Childress, 2001; Lee *et al.,* 2002; Tanninen *et al.,* 2002; Yoon *et al.,* 2002), which results in increased electrostatic repulsion between a negatively charged solute and membrane and consequently rejects more solute from water. Conversely, it was determined that the presence of counter ions (Na^+, K^+, Ca^{2+}, and Mg^{2+}) in feed water can decrease the membrane rejection of negatively charged solute (Braghetta *et al.,* 1997; Ariza *et al.,* 2002; Yoon *et al.,* 2002).

However, the influence of pH and membrane surface charge on membrane pore structure and the rejection of uncharged organics as well as permeate flux is somewhat contradictory (Berg *et al.,* 1997; Braghetta *et al.,* 1997; Yoon *et al.,* 1998; Childress and Elimelech, 2000; Freger *et al.,* 2000; Boussahel *et al.,* 2002; Lee *et al.,* 2002; Ozaki and Li, 2002).

Since most high-pressure membranes are considered hydrophobic, the adsorption of hydrophobic compounds onto membranes may be an important factor in the rejection of micropollutants during membrane applications. In fact, many studies have conformed that hydrophobic–hydrophobic interactions between solute and membrane are an important factor for the rejection of hydrophobic compounds and that steric hindrance may also contribute to rejection (Kiso *et al.,* 2001a; Nghiem *et al.,* 2004; Van der Bruggen *et al.,* 2001a; Van der Bruggen *et al.,* 2001b; Nghiem *et al.,* 2002; Van der Bruggen *et al.,* 2002a; Agenson *et al.,* 2003; Kimura *et al.,* 2003a; Wintgens *et al.,* 2003). Among those studies, several parameters such as octanol-water distribution coefficient (K_{ow}), Taft and Hammett numbers (effect of the substituent group on polarity) and Dvs (measure of the stretching of the OH bond) were found to correlate with the rejection of these compounds.

Verliefde *et al.* (2007) offered a qualitative prediction of nanofiltration rejection for the selected priority micropollutants in Flemish and Dutch water sources. The qualitative prediction was based on the values of key solute and membrane parameters in nanofiltration. The qualitative predictions are roughly in agreement with literature values and may provide very quick and useful technique to assess the implementation of nanofiltration as a treatment step for organic micropollutants in drinking water plant design.

Feed water composition can certainly have a significant effect upon adsorption effects and rejection. Many studies have reported that the complexity of rejection mechanism and the effect feed water composition might have on solute rejection. (Tödtheide *et al.,* 1997; Kiso *et al.,* 2001a; Majewska-Nowak *et al.,* 2002; Schäfer *et al.,* 2002).

Fouling is one of the main problems in any membrane separation. Issues associated with membrane fouling remain quite problematic with respect to not only volume production but also permeate quality. Fouling may occur in pores of membranes by partial pore size reduction caused by foulants adsorbing on the inner pore walls, pore blockage and surface fouling such as cake and gel layer formation. The presence of the fouling layer can drastically alter the characteristics of the membrane surface including surface charge and hydrophobicity (Childress and Elimelech, 2000; Xu *et al.,* 2006). Consequently, in addition to a reduction of flux, membrane fouling can lead to considerable variation in the membrane separation efficiency. It has been reported that membrane fouling can either improve or deteriorate permeate quality but the negative consequences of fouling are obvious including the need for pretreatment, membrane cleaning, limited recoveries and feed water loss, and short lifetimes of membranes. A wide spectrum of constituents in process waters contribute to fouling. These include inorganic solutes, dissolved and macromolecular organic, suspended particles and biological solids.

Scaling usually refers to the formation of deposits of inverse-solubility salts. The greatest scaling potential species in NF membrane are $CaCO_3$, $CaSO_4 \cdot 2H_2O$ and silica, while the other potential scaling species are $BaSO_4$, $SrSO_4$, $Ca(PO4)_2$, ferric and aluminium hydroxides (Faller, 1999; Al-Amoudi and Lovitt, 2007). Inorganic scale formation can even lead to physical damage of the NF membrane, and it is difficult to restore NF membrane performance due to the difficulties of scale removal and irreversible membrane pore plugging (Jarusutthirak *et al.,* 2002; Al-Amoudi and Lovitt, 2007).

Organic fouling can be influenced by: membrane characteristics (Elimelech *et al.,* 1997; Schäfer *et al.,* 1998; Van der Bruggen *et al.,* 1999; Mänttäri, 2000;

Van der Bruggen *et al.,* 2002b), including surface structure as well as surface chemical properties, chemistry of feed solution including ionic strength (Ghosh and Schnitzer, 1980; Elimelech *et al.,* 1997), pH (Childress and Elimelech, 1996; Childress and Deshmukh, 1998; Schäfer *et al.,* 1998; Mänttäri, 2000; Schäfer *et al.,* 2004); the concentration of monovalent ions and divalent ions (Elimelech *et al.,* 1997; Schäfer *et al.,* 1998, 2004); the properties of NOM, including molecular weight and polarity (Van der Bruggen *et al.,* 1999, 2002b; Bellona *et al.,* 2004); the hydrodynamics and the operating conditions at the membrane surface including permeate flux (Van der Bruggen *et al.,* 2002b), pressure (Schäfer *et al.,* 1998; Le Roux *et al.,* 2005), concentration polarization (Schäfer *et al.,* 1998), and the mass transfer properties of the fluid boundary layer.

Organic fouling could cause either reversible or irreversible flux decline. The reversible flux decline, due to NOM fouling, can be restored partially or fully by chemical cleaning (Al-Amoudi and Farooque, 2005). However, the irreversible flux decline can not be restored at all even by rigorous chemical cleaning is applied to remove NOM (Roudman and DiGiano, 2000).

Different types of fouling may occur simultaneously and can influence each other (Flemming, 1993). Scaling by inorganic compounds is usually controlled using a scale inhibitor, such as a polymer or an acid. Particulate fouling can be controlled by pretreatment, such as ultrafiltration. Thus, all types of fouling except biofouling and organic fouling – likely related types of fouling – are controllable. Numerous authors describe biofouling problems in membrane installations (Flemming, 1993; Tasaka *et al.,* 1994; Ridgway and Flemming, 1996; Baker and Dudley, 1998; Schneider *et al.,* 2005; Karime *et al.,* 2008).

Biofouling is hard to quantify because no univocal quantification methods related to biofouling and operational problems are described. Pressure drop is generally used as a good parameter for evaluating fouling. An increase of pressure drop is, however, not conclusively linked to biofouling, since other factors may influence the pressure drop as well. Additionally, the pressure drop measurement may not be sensitive enough for early detection of biofouling. Biofouling can be controlled by (1) removal of degradable components from the feed water, (2) ensuring the relative purity of the chemicals dosed and (3) performing effective cleaning procedures (Ridgway and Flemming, 1996; Baker and Dudley, 1998; Jarusutthirak *et al.,* 2002).

A quick scan of membrane fouling is necessary to give the conclusive information about the types and extent of fouling as well as the fouling control in the membrane filtration process.

The Silt Density Index (SDI) and Fouling Index (FI) are presently used to measure the colloidal fouling potential of feedwater, but their limitations have been evidenced by several studies (Schippers and Verdouw, 1980; Boerlage *et al.*, 2003). Schippers and Verdouw (1980) developed the $MFI_{0.45}$ which was unable to take into account the influence of the colloidal particles. Boerlage *et al.* (1997) proposed a MFI–UF using a polyacrylonitrile membrane (PAN) with a MWCO of 13 kDa as a reference membrane and to measure the fouling potential of the feed water. The MFI-UF was found to be a promising tool for measuring the colloidal fouling potential for RO NF and UF systems. Roorda and Van der Graaf (2001) defined Normalized MFI–UF and proposed to give the results under standard conditions (1 m^2 membrane area and 1 bar trans-membrane pressure). Rabie *et al.* (2001) developed a method for the optimization of long-term operation of a membrane unit using the analysis of initial performance. However, the results obtained from Brauns *et al.* (2002) revealed that FI should be used as an intrinsic character of water but not as a parameter for design purposes.

As mentioned as above, MF and UF membranes are currently used for MFI. However, a fraction of colloids and solutes are not retained by these membranes and thus are not taken in account. For example, natural organic matter (NOM) and and even more of effluent organic matter (EfOM) contain a fraction of organics which molecular weight (MW) is around 1000Da (Abdessemed *et al.*, 2002; Jarusutthirak *et al.*, 2002).

Khirani *et al.* (2006) employed a loose, uncharged NF membrane with a MWCO of 500–1500 Daltons (hydrophylic polyether sulphone membranes with a thin-film oxidation resistant layer) to develop a NF-MFI. The limiting parameter of the MFI–UF is the duration of the test (more than 20 h). Khirani's method showed that the determination of a FI was possible in a short time (about 1 h) using a loose NF or a NF membrane. The hypothesis of Khirani's study is that a NF membrane is able to retain all the components responsible for fouling including small molecules (colloids and solutes) that are involved in membrane fouling. Obviously, the study showed that dissolved organics are responsible for fouling and have to be taken into account in determining the membrane fouling potential of feed water.

In addition to MFI, several methods developed for the diagnosis, prediction, prevention, and control of fouling have been proposed and applied in practice and have proven their value in controlling fouling. An overview of the coherent tools is shown in Table 4.1 (Vrouwenvelder *et al.*, 2003).

Table 4.1 Overview of tools available for determining the fouling potential of feed water and fouling diagnosis of NF and RO membranes used in water treatment (Vrouwenvelder *et al.*, 2003)

Tools	Fouling diagnosis	Comment
Integrated diagnosis (autopsy)	Biofouling, inorganic, compounds and particles	Diagnosis of foulant in membrane elements
Biofilm monitor and AOC	Biofouling	Predictive and prevention of biofouling by determining the (growth) potential of water
SOCR	Biofouling	Non-destructive method for determining active biomass in membrane systems
MFI-UF	Particulate	Particulate fouling potential of water
ScaleGuard	Scaling	Optimizing recovery, acid dose and anti-scalant dose

4.4 NANOFILTRATION OF MICROPOLLUTANTS IN WATER

Nanofiltration membranes were initially developed for softening purposes. Nowadays, although softening is still a major application, nanofiltration is a rapidly developing technology with promising applications to remove pesticides and other organic contaminants from surface and ground waters to help insure the safety of public drinking water supplies.

Softening is a typical process for groundwater treatment. The traditional methods for water softening include lime-soda and ion exchange processes. In contact with an aqueous solution, most NF membranes become positively or negatively charged due to the presence of ionizable groups. Therefore, NF membranes can also be used to remove small ionic components or inorganic salts. The softening of ground water using NF has been studied by many investigators.

A comparison between lime softening and nanofiltration for groundwater treatment in Florida has been carried out by Bergman (1995). Cost evaluation was done under several operating modes in the study. The cost of membrane softening can be even lower than for lime softening if additional treatment processes are added to lime softening to match the better membrane softening permeate quality, or if some water can be bypassed around the membranes and blended to produce water comparable to the finished water in the lime softening plant. Definitely, NF membrane softening is becoming an attractive alternative presenting many advantages: superior water product quality, no sludge disposal, ease of operation as well as reduction of overall plant construction and O&M costs.

Sombekke *et al.* (1997) made a comparison between nanofiltration and pellet softening (combined with granular activated carbon (GAC) adsorption for organics removal) based on a life cycle analysis (LCA). The environmental impact of a product in its entire life cycle and all the extractions from and emissions into the environment were involved in the LCA study. Both treatment schemes were found to have a comparable impact. However, nanofiltration was advantageous for quality and health aspects.

The need for partial softening of raw waters as well as the removal of organic micropollutants, has led to the adaptation of nanofiltration for serving this dual purpose. An example is the treatment of water from a lake in Taiwan (Yeh *et al.*, 2000), where hardness, taste and odor problems had to be solved at the same time. Yeh *et al.* used different methods such as a conventional process followed by ozone, GAC and pellet softening, and an integrated process of membrane process (UF/NF) and conventional process. Softening was satisfactory for all processes, but water produced by the membrane process had the best quality as measured by turbidity, dissolved organics, biostability and organoleptic parameters.

The development of membrane with high rejection of organics but low hardness rejection was carried out in 1997 (De Witte, 1997). De Witte demonstrated that NF 200 (Filmtec) membrane performance remains good after repetitive cleaning and energy consumption is still low. NF 200 (Filmtec) membrane was successfully used at Debden Road water works, Saffron Walden, England (Wittmann *et al.*, 1998).

Fu *et al.* (1994, 1995) used NTR 7450 membrane made by Nitto-Denko to remove organics without removing much of the inorganics and the permeability of the membrane were superior compared to traditional NF membranes used for softening. The NTR 7450 could be operated at a recovery of 90% and a flux of 34 l/m^2h and organics removal was nearly complete.

Nanofiltration is also considered as an alternative that can be used to meet regulations for lowered arsenic concentrations in drinking water (Kartinen and Martin, 1995). Saitúa *et al.* (2005) reported that arsenic rejection by nanofiltration was independent of transmembrane pressure, crossflow velocity and temperature. The co-occurrence of dissolved inorganics does not significantly influence arsenic rejection. Waypa *et al.* (1997) studied the arsenic removal from synthetic freshwater and from surface water sources by NF. They presented that both As(V) and As(III) were effectively removed from the water by NF membrane over a range of operating conditions. The NF membrane can achieve rejection of 99%. The result also showed that the removal of As(V) and As(III) was comparable, with no preferential rejection of As(V) over As(III). The authors have concluded that size exclusion governed their separation behavior and not the charge interaction. Seidel *et al.* (2001) studied the difference in rejection between As(V) and As(III) using loose (porous) NF membranes. The removal of As(V) was varied between 60% and 90%, whereas As(III) was below 30%. Sato *et al.* (2002) also investigated the performance of nanofiltration for arsenic removal. In their studies, NF membranes could remove over 95% of pentavalent arsenic and more than 75% of trivalent arsenic could be removed without any chemical additives. Furthermore, both As(V) and As(III) removal by NF membranes was not affected by source water chemical compositions. An overview on the removal of arsenic from surface water and ground water by nanofiltration of is also reported by Van der Bruggen and Vandecasteele (2003).

Gestel *et al.* (2002b) employed a multilayer TiO_2 membrane for the retention of five types of salts. The membrane showed aminimal salt retention at pH 6 and a fairly high retention at alkaline pH (R(NaCl) = 85%, R(KCl) = 87%, R(LiCl) = 90%). For salts containing divalent ions, high retentions were again achieved (R(Na2SO4) >95%; R(CaCl2) = 78%).

El-Sheikh *et al.* (2007) used different kinds of multi-walled carbon nanotube (MWCNT) for enrichment of metal ions (Pb^{2+}, Cd^{2+}, Cu^{2+}, Zn^{2+} and MnO_4^-) from environmental waters prior to their analysis. It was found that long MWCNT of length 5–15 μm and external diameter 10–30 nm gave the highest enrichment efficiency towards MnO_4^-, Cu^{2+}, Zn^{2+} and Pb^{2+}; but not for Cd^{2+} due to its low recovery.

Lin and Murad (2001b) reported that 100% Na^+ rejection could be achieved on a perfect (single crystal), ZK-4 membrane through RO. The separation mechanism of the perfect ZK-4 zeolite membranes is the size exclusion of hydrated ions, which have kinetic sizes significantly larger than the aperture of the ZK-4 zeolite.

Li *et al.* (2004a, b) used MFI-type zeolite membranes in RO separation and showed 77% rejection of Na^+. In a complex feed solution containing 0.1M NaCl + 0.1M KCl + 0.1M NH_4Cl + 0.1M $CaCl_2$ + 0.1M $MgCl_2$, rejections of Na^+, K^+, NH_4^+, Ca^{2+}, and Mg^{2+} were 58.1%, 62.6%, 79.9%, 80.7%, and 88.4%, respectively.

Choi *et al.* (2008) investigated the effect of co-existing ions on the removal of several anions using negative surface charge nanomembrane. The results showed that sulfate would be rejected most among other ions in the groundwater. The chloride ions were rejected more than nitrate and fluoride ions. The experiment indicated that the electric repulsion between the nanomembrane and chloride ion was so high that it could even push some divalent sulfate ions through the membrane. Fluoride was less affected by the surface charge of the membrane than nitrate. The hydration effect of nitrate would be stronger at a membrane with lower surface potential. The experiment indicated that calcium ions shielded membrane charges more effectively than magnesium ions. However, despite the charge shielding effect, the rejection rates against the divalent anion were high, and more ions were rejected by the membrane that has a high negative surface potential.

Removal of natural organic matter (NOM) and disinfection byproduct (DBP) precursors from water sources by NF membranes have been studied by many investigators (Agbekodo *et al.*, 1996; Ericsson *et al.*, 1996; Alborzfar *et al.*, 1998; Visvanathan *et al.*, 1998; Cho *et al.*, 1999; Levine *et al.*, 1999; Escobar *et al.*, 2000; Everest and Malloy, 2000; Khalik and Praptowidodo, 2000). It is obvious that the best results were obtained with membranes with a MWCO around 200.

Visvanathan *et al.* (1998) evaluated the effects of interference parameters include operating pressure, feed THMPs concentration, pH, presence of other ions (Ca^{2+} and Mg^{2+}), and suspended solids on the performance of nanofiltration for removal of trihalomethane precursors (THMPs). Generally rejection was found to be greater than 90% for a precompacted membrane. Experimental results also showed that higher pressure, feed THMP concentration, and suspended solids increased rejection and divalent ions reduced the rejection capacity.

NOM essentially consists of molecules from a large range of molecular weights, this confirms the need for a membrane with a low MWCO for complete organics removal. Agbekodo *et al.* (1996a) reported that about 60% of the remaining DOC in NF permeate is caused by amino acids as well as lower fractions of fatty-aromatic acids and aldehydes.

The removal of anthropogenic micropollutants from water by NF membrane has been studied by many researchers (Agbekodo *et al.*, 1996b; Montovay *et al.*, 1996; Van der Bruggen *et al.*, 1998; Ducom and Cabassud, 1999; Kiso *et al.*, 2000; Kimura *et al.*, 2003; Causserand *et al.*, 2005; Plakas *et al.*, 2006; Lee *et al.*, 2008). Since the majority of the compounds categorized as pesticides have molecular weights (MW) of more than 200 Da, nanofiltration (NF) seems to be a promising option for their removal from contaminated water sources. However, the results came from those studies showed the removal efficiencies largely depend on the membranes used and on the micropollutants that have to be removed.

Agbekodo *et al.* (1996b) demonstrated the influence of natural organic matter (NOM) on the retention of atrazine and simazine. The removal efficiency of the NF70 membranes can vary from 50 to 100% as a function of the DOC level in the feed water. Montovay *et al.* (1996) found an 80% removal of atrazine and a 40% removal of metazachlor, which is insufficient. Van der Bruggen *et al.* (1998) showed that the NF70 membrane can reject pesticides such as atrazine, simazine, diuron and isoproturon over 90%. However, relatively low rejections were found for diuron and isoproturon for two other membranes (NF45 and UTC-20). Ducom and Cabassud (1999) studied the removal of trichloroethylene, tetrachloroethylene and chloroform by nanofiltration. The removal of trichloroethylene and tetrachloroethylene can be achieved using several different types of NF membranes, but chloroform rejection was significantly lower. However, good removal efficiencies of chloroform were obtained by Waniek *et al.* (2002). Kiso *et al.* (2000) studied the removal of 12 pesticides. Rejections obtained with three of these membranes were too low; the rejections with the fourth membrane were very high (over 95%), but this membrane appears to be a reverse osmosis membrane, given the high NaCl rejection. Kimura *et al.* (2003) studied the rejection of disinfection by-products (DBPs), endocrine disrupting compounds (EDCs), and pharmaceutically active compounds (PhACs) by nanofiltration (NF) and reverse osmosis (RO) membranes as a function of their physico-chemical properties and initial feed water concentration. Experimental results indicated that negatively charged compounds could be rejected very effectively (i.e., $>90\%$) regardless of other physico-chemical properties of the tested compounds and rejection of the compounds were not time-dependency. Contrarily, rejection of non-charged compounds was generally lower ($<90\%$ except for one case) and influenced mainly by the molecular size of the compounds. A clear time-dependency was observed for rejection of non-charged compounds, attributable to compound adsorption on the membrane. Causserand *et al.* (2005) reported the performance of polyamide membrane was much better that that of cellulose acetate membrane in removing 2,4-dichloroaniline. Plakas *et al.* (2006) investigated the role of organic matter and calcium concentration on the removals of atrazine, isoproturon and prometryn. The results showed that nanofiltration of water where herbicides are present together with humic substances results in increased herbicide retention. This trend is less evident in the presence of calcium ions due to their possible interference with the humic substances–herbicides interactions. Lee *et al.* (2008) demonstrated the effects of membrane properties and solution chemistries on removal efficiencies of TCEP and perchlorate.

Basically, the size of bacteria (0.5–10 μm) and protozoan cysts and oocysts (3–15 μm) are larger than the pore size of UF membranes. Both of them can be removed with at least 4 log units using UF membranes. The size of a virus varies between 20 and 80 nm, whereas UF membranes have pores of approximately 10 nm and more, so that complete removal of virus by UF membranes is theoretically possible. For NF membranes which have pore sizes below 1 nm, the smaller viruses may be rejected. In fact, NF membranes were found to be able to remove viruses and bacteria from surface water quite successfully [11,38–41].

Yahya *et al.* (1993) compared the performances of slow sand filtration and NF membranes in a 76 m^3/d surface water pilot plant for the removal of two bacteriophages (MS-2, 28 nm, and PRD-1, 65 nm). The slow sand filters removed

99 and 99.9% of the bacteriophages, respectively, and the NF membranes (NF70 – Filmtec; Desal 5 DK and Desal 5 SG–Osmonics) removed 4–6 log units of the test viruses. Otaki *et al.* (1998) reported a log 7 removal of poliomyelitis virus vaccine and a log 6 removal of coliphage Q beta from river water in the Tokyo area by the NTR-729HFS4 membrane (Nitto-Denko). Reiss *et al.* (1999) used an integrated membrane system of microfiltration with NF to remove *Bacillus subillus* spores from 5.4 to 10.7 log. Urase *et al.* (1996) used MF, UF and NF membranes for removal of the model viruses Q beta and T4. The removal of the test viruses ranged from 2 to 6 log units through different membranes. It can be concluded that 100% virus retention cannot be obtained with pressure driven membrane processes due to the leakage of viruses through 'abnormally large' pores.

Brady-Estevez *et al.* (2008) demonstrate the use of a single-walled carbon nanotube (SWNT) for the effective removal of bacterial and viral pathogens from water at low pressures. The filter was developed using a poly(vinylidene fluoride) (PVDF)-based microporous membrane (5 mm pore size) covered with a thin layer of SWNTs. Such a hybrid filter would be exceptionally robust, permitting reuse, as the high thermal resistance of carbon nanotubes and ceramics would allow for simple thermal regeneration of the filter. For bacterium removal study, Escherichia coli K12 was selected as a model bacterium. The study showed that viruses can be completely removed by a depth-filtration mechanism (Figure 4.4), that is, capture by nanotube bundles inside the SWNT layer. For virus rejection study, a model virus particle, MS2 bacteriophage, diameter 27 nm, was selected. The results showed that virus removal from the 10^7 virus particles per mL initial concentration was complete, without any viral particles detected by the PFU (plaque forming unit) method at the filter outlet.

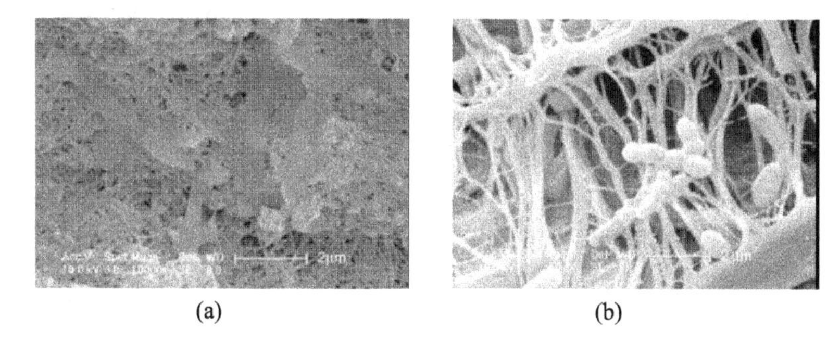

Figure 4.4 Retention of E. coli by SWNT filter. (a) SEM image of E. coli cells retained on SWNT filter. (b) SEM image of E. coli cells on the base membrane (5 mm pore PVDF membrane)

Normally, NF is not thought of as a pure disinfection process and post-treatment of the NF permeate will always be necessary. Moreover, chlorination might be required for the NF permeate, but would likely be low in DBPs formation, to prevent bacterial regrowth in the distribution network (Laurent *et al.,* 1999). Another alternative is to integrate with RO for the further improvement of disinfection.

A case study at Lake Arrowhead, California using a combination of coagulation/flocculation/sedimentation, sand filtration, ozonation (in two stages), GAC filtration, UF/NF, and RO showed that the integrated system can remove 21–22 log units of bacteriophage and 8–10 log units of Giardia and Cryptosporidium (Madireddi *et al.,* 1997).

Among MF, UF, NF and RO membranes, NF in particular offers a comprehensive approach to meeting multiple water quality objectives including removal of dissolved organics and inorganic contaminants. In addition, NF has obvious advantages compared to RO mainly: (1) lower operating pressure and (2) selective rejection between monovalent and multivalent ions.

Feed pretreatment is one of the major factors determining the success or failure of a desalination process. Pre-treatment of seawater feed to RO/thermal processes using nanofiltration is expected to lower the required pressure to operate RO plant by reducing seawater feed TDS as well as to reduce the energy the energy consumption (Redondo, 2001).

The main drawback of conventional pretreatment based on chemical and mechanical treatments (i.e., coagulation, flocculation, acid treatment, pH adjustment, addition of anti-sealant and mediafiltration) is known to be complex, labor intensive and space consuming (Sikora *et al.,* 1989; Van Hoop *et al.,* 2001). Another problem in using conventional pretreatment is corrosion and corrosion products since the acid dosing system is commonly used in the conventional process (Sikora *et al.,* 1989; AI-Ahmad and Adbul Aleem, 1993).

NF was used for the first time by Hassan *et al.* (1998) as pretreatment for seawater reverse osmosis (SWRO), multistage flash (MSF), and seawater reverse osmosis rejected in multistage flash ($SWRO_{rejected}$-MSF) processes. The NF applicaton made it possible to operate a SWRO and MSF pilot plant at a high recovery of 70% and 80%, respectively.

Regarding the feed water quality improvement of SWRO desalination plants, NF pretreatment can be beneficial to (1) prevent SWRO membrane fouling by the removal of turbidity and bacteria, (2) prevent scaling (both in SWRO and MSF) by removal of scale forming hardness ions, and (3) lower required pressure to operate SWRO plants by reducing seawater feed TDS by 30–60%, depending on the type of NF membrane and operating conditions (Al-Sofi *et al.,* 1998; Cfiscuoli and Drioli, 1999; Hassan *et al.,* 2000; Al-Sofi, 2001; Drioli *et al.,* 2002; Mohesn *et al.,* 2003; Pontié *et al.,* 2003).

A promising approach for pretreatment of seawater make-up feed to MSF and SWRO desalination processes using nanofiltration (NF) membranes has been introduced by the R&D Center (RDC) of SWCC. NF membranes are capable to reduce significantly scale forming ions from seawater, allow high temperature operation of thermal desalination processes, and subsequently increase water productivity (Hamed, 2005).

Hafiarle *et al.* (2000) used a TFC-S NF membrane to remove chromate from an aqueous solution. The results showed that the rejection depended on the ionic strength and pH. Better retention was obtained at basic pH (up to 80% at a pH of 8). Results also showed that NF is a very promising method of treatment for wastewater charged with hexavalent chromium.

Ku *et al.* (2005) studied the effect of solution composition on the removal of copper ions by nanofiltration. The results indicated that the rejection of copper ions increases with increasing the charge valence of co-anions present in aqueous solution. The surfactant presented in aqueous solution was adsorbed by the membrane to form a secondary filtration layer on the membrane surface, therefore, influenced the surface charge characteristic of the membrane.

Choi *et al.* (2006) evaluated the application potential of nanofiltration membranes for the rejection of organic acids in wastewaters. The rejection of succinic and citric acids, which have molecular weights (M_Ws) larger than or closer to the molecular weight cutoffs (MWCOs) of employed NF membranes, was over 90% irrespective of operating pressure. Contrarily, the rejection of organic acids with M_Ws much smaller than MWCOs of the NF membranes increased gradually with increasing the applied pressure. The increase of DOC rejection with filtration time could be explained by increase of electrostatic repulsion between membrane and dissociated organic acid and by membrane fouling. NF process showed considerable potential as an advanced wastewater treatment process for removing organic acids in wastewaters.

Kim *et al.* (2007) reported the nitrate rejection of nanomembrane (NTR 729HF) from stainless steel industry. The results showed that the rejection rate of NF was decreased as pH decreased and Ca_2^+ concentration increased indicating that charge repulsion is one of the major rejection mechanisms.

Ortega *et al.* (2008) used two commercial nanofiltration membranes to remove metal ions from an acidic leachate solution generated from a contaminated soil using H_2SO_4 as a soil washing agent. Two types of thin-film commercial NF membranes, Desal5 DK and NF-270, were studied for their permeation and ionic selectivity. Desal5 DK is a polymeric membrane in which a polyamide selective layer is supported on a polysulfone layer. NF-270 is a semi-aromatic piperazine-based polyamide layer on top of a polysulphone micro-porous support reinforced with a polyester non-vowen backing layer. The results demonstrated the effectiveness and feasibility of the application of nanofiltration treatments in the cleaning-up of contaminated water residues generated during soil washing processes.

The application of high temperature resistant membranes is increasingly gaining attention in industry because they have many advantages comparing to most commercial polymer NF membranes which can only be applied under 45–50°C. High temperature resistant membranes can be used in the treatment of various hot fluid streams without strict temperature control. What is more, the enhanced flux due to high temperature may allow certain reduction of operating pressure, which further saves operating cost.

Tang and Chen (2002) used NF to treat textile wastewater which was highly colored with a high loading of inorganic salts. The results showed that the rejection of dye was 98% under an operating pressure of 500 kPa, and the NaC1 rejection was less than 14%. Results also showed that NF is a very promising method for water reuse of textile industries.

Wu *et al.* (2009) prepared a novel thermal stable composite NF membrane by interfacial polymerization of piperazine (PIP) and trimesoyl chloride (TMC) on the substrate of thermal stable PPEA UF membrane. The purification experiments were accomplished effectively with a rejection of 99.3% for dyes Congo red (CGR) and Acid chrome blue K (ACBK) at 1.0 MPa, 80°C.

Voigt *et al.* (2001) reported a new TiO_2-NF ceramic membrane to decolor textile wastewater using an integrated pilot plan. The results show that it is possible to treat textile wastewater with dye retention varying from 70% to 100%, COD reduction of 45–80%, and salt retention of 10–80%.

Reverse osmosis is an effective technology to remove organic compounds from water bodies, especially for those that contain low concentration and low molecular weight organic compounds. Traditional RO membrane is limited due to high operational cost and maintenance as RO involves requirement of high pressure to the system and need extensive pretreatment. In addition, more and more current water and wastewater treatment plants are requested to higher their water recovery rate, which should be close to 100%. To overcome the limitations of RO, many researchers investigated an integrated membrane system (IMS) and many evidences indicated that NF membranes could be an alternative for the integration system (Nederlof *et al.*, 2000; Huiting *et al.,* 2001; Kimura *et al.,* 2003b; Zhao *et al.,* 2005; Bellona *et al.,* 2007; Jacob *et al.,* 2009; Simon *et al.,* 2009). Since NF membranes are generally supplied in the same configurations as RO membranes, utilities could replace RO with NF spiral-wound elements without the need for significant additional capital investment.

A general schematic plot of an integrated membrane system (IMS) for tap-water supply designed and installed by KINTECH Technology Co. Ltd. is depicted in Figure 4.5. The plant located in southern Taiwan was established by Taiwan Water Corporation in 1972 and initially operated on conventional process consisted of coagulation, flocculation, setting, air stripping and sand filtration. To meet new water quality requirement, the plant was upgraded in 2007 by integrating IMS (UF-NF/RO, Figure 4.6) into the conventional system. A flexible operating mode was adopted by mixing sand filtration treated water or UF permeate with RO permeate to save the operating cost and meet the water quality standard items of turbidity (turb. < 0.2 mg/L), total hardness (TH < 150 mg/L as $CaCO_3$), total dissolved solid (TDS < 250 mg/L), *Escherichia coli* (E. coli = 0.0 CFU/100ml) and total trihalomethanes (TTHMs < 30μg/L). The full capacity of this plant is 303,400 m^3/day includes 170,000 m^3/day of integrated membrane system. The total land area of IMS is 1380 m^2 and the recovery of IMS is 90%. Table 4.2 shows the raw and treated water quality of Caotan water purification plant.

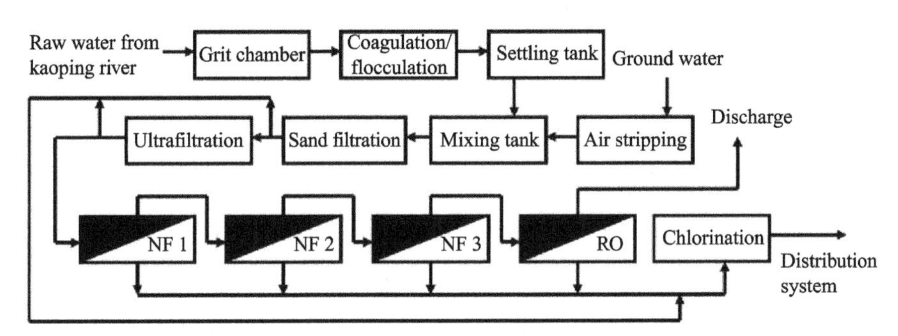

Figure 4.5 Schematic Flow Diagram of Caotan water purification plant showing the arrangement of UF, NF and RO membranes

Figure 4.6 Full-scale integrated membrane system (IMS) of Caotan water purification plant for tap water supply located at Kaohsiung County, Taiwan (a) Ultrafiltration system; (b) NF-LPRO system. The total land area of IMS is 1380 m^2. (Photos provided by KINTECH Technology Co., Ltd.)

Table 4.2 Water quality of Caotan water purification plant

	River water	Ground water	RO permeate	Mixed treated water after chlorination
Turbidity (mg/L)	15–15000	2–5300	–	0.11
TH (mg/L as $CaCO_3$)	190–310	320–480	20	135
TDS (mg/L)	260–550	470–680	–	240
E. coli (CFU/100ml)	–	–	–	0
TTHMs (μg/L)	–	–	–	10

4.5 REFERENCES

Abdessemed, D., Nezzal G. and Ben Aim, R. (2002) Fractionation of a secondary effluent with membrane separation. *Desalination* **146**(1–3), 433–437.

Agbekodo, K. M., Legube B. and Cote, P. (1996a) Organics in NF permeate. *J. AWWA* **88**(5), 67–74.

Agbekodo, K. M., Legube, B. and Dard, S. (1996b) Atrazine and simazine removal mechanisms by nanofiltration: Influence of natural organic matter concentration. *Water Res.* **30**(11) 2535–2542.

Agenson, K. O., Oh, J.-H. and Urase, T. (2003) Retention of a wide variety of organic pollutants by different nanofiltration/reverse osmosis membranes: controlling parameters of process. *J. Membr. Sci.* **225**(1–2), 91–103.

Agnihotri, S., Rood, M. J. and Rostam-Abadi, M. (2005) Adsorption equilibrium of organic vapors on single-walled carbon nanotubes. *Carbon* **43**(11), 2379–2388.

Al-Ahmad, M. and Adbul Aleem, F. (1993) Scale formation and fouling problems effect on the performance of MSF and RO desalination, plants in Saudi Arabia. *Desalination* **93**(1–3), 287–310.

Alami-Younssi, S., Larbot, A., Persin, M., Sarrazin, J. and Cot, L. (1995) Rejection of mineral salts on a gamma alumina nanofiltration membrane: Application to environmental processes. *J. Membr. Sci.* **102**, 123–129.

Al-Amoudi, A. and Lovitt, R. W. (2007) Fouling strategies and the cleaning system of NF membranes and factors affecting cleaning efficiency. *J. Membr. Sci.* 303(1–2), 4–28.

Al-Amoudi, A. S. and Farooque, A. M. (2005) Performance, restoration and autopsy of NF membranes used in seawater pretreatment. *Desalination* **178**(1–3), 261–271.

Alborzfar, M., Escande, K. and Allen, S. J. (1998) Removal of natural organic matter from two types of humic ground waters by nanofiltration. *Water Res.* **32**(10), 2970–2983.

Al-Sofi, M. A. K., Hassan, A. M., Mustafa, G. M., Dalvi, A. G. I. and Kither, M. N. M. (1998) Nanofiltration as a means of achieving higher TBT of ⩾120 degrees C in MSF. *Desalination* **118**(1–3), 123–129.

Al-Sofi, M. A.-K. (2001) Seawater desalination – SWCC experience and vision. *Desalination* **135**(1–3), 121–139.

Ariza, M. J., Canas, A., Malfeito, J. and Benavente, J. (2002) Effect of pH on electrokinetic and electrochemical parameters of both sub-layers of composite polyamide/polysulfone membranes. *Desalination* **148**(1–3), 377–382.
Baker, J. S. and Dudley, L. Y. (1998) Biofouling in membrane systems – a review. *Desalination* **118**(1–3), 81–90.
Baticle, P., Kiefer, C., Lakhchaf, N., Larbot, A., Leclerc, O., Persin, M. and Sarrazin, J. (1997) Salt filtration on gamma alumina nanofiltration membranes fired at two different temperatures. *J. Membr. Sci.* **135**(1), 1–8.
Bellona, C. and Drewes, J. E. (2007) Viability of a low-pressure nanofilter in treating recycled water for water reuse applications: A pilot-scale study. *Water Res.* **41**(17), 3948–3958.
Bellona, C., Drewes, J. E., Xu, P. and Amy, G. (2004) Factors affecting the rejection of organic solutes during NF/RO treatment – a literature review. *Water Res.* **38**(12), 2795–2809.
Berg, P., Hagmeyer, G. and Gimbel, R. (1997) Removal of pesticides and other micropollutants by nanofiltration. *Desalination* **113**(2–3), 205–208.
Bergman, R. A. (1995) Membrane softening versus lime softening in Florida-a cost comparison update. *Desalination* **102**(1–3), 11–24.
Boeflage, S., Kennedy, M., Bonne, P. A. C., Galjaard G. and Schippers, J. (1997) Prediction of flux decline in membrane systems due to particulate fouling. *Desalination* **113**(2–3), 231–233.
Boerlage, S. F. E., Kennedy, M., Aniye, M. P. and Schippers, J. C. (2003) Applications of the MFI-UF to measure and predict particulate fouling in RO systems. *J. Membr. Sci.* **220**(1–2), 97–116.
Boussahel, R., Montiel, A. and Baudu, M. (2002) Effects of organic and inorganic matter on pesticide rejection by nanofiltration. *Desalination* **145**(1–3), 109–114.
Bowen, W. R. and Mohammad A. W. (1998) Diafiltration by nanofiltration: prediction and optimization. *AIChE J* **44**(8), 1799–1811.
Bowen, W. R., Welfoot, J. S. and Williams, M. (2002) Linearized transport model for nanofiltration: Development and assessment. *AIChE J* **48**(4), 760–771.
Brady-Estévez, A. S., Kang, S. and Elimelech, M. (2008) A single-walled-carbon-nanotube filter for removal of viral and bacterial pathogens. *Small* **4**(4), 481–484.
Braghetta, A., Digiano, F. A. and Ball, W. P. (1997) Nanofiltration of natural organic matter: pH and ionic strength effects. *J. Environ. Eng.* **123**(7), 628–640.
Brandhuber, P. and Amy, G. (1998) Alternative methods for membrane filtration of arsenic from drinking water. *Desalination* **117**(1–3), 1–10.
Brauns, E., Van Hoof, E., Molenberghs, B., Dotremont, C., Doyen, W. and Leysen, R. (2002) A new method of measuring and presenting the membrane fouling potential. *Desalination* **150**(1), 31–43.
Burggraaf, A. J. and Keizer, K. (1991) Synthesis of inorganic membranes. In Inorganic Membranes: Characterization and Applications (ed. Bhave, R. R.), Van Nostrand Rheinhold, New York, pp. 10–63.
Causserand, C., Aimar, P., Cravedi, J. P. and Singlande, E. (2005) Dichloroaniline retention by nanofiltration membranes. *Water Res.* **39**(8), 1594–1600.
Cfiscuoli A. and Drioli, E. (1999) Energetic and exergetic analysis of an integrated membrane desalination system. *Desalination* **124**(1–3), 243–249.
Childress, A. E. and Deshmukh, S. S. (1998) Effect of humic substances and anionic surfactants on the surface charge and performance of reverse osmosis membranes. *Desalination* **118**(1–3), 167–174.

Childress, A. E. and Elimelech, M. (1996) Effect of solution chemistry on the surface charge of polymeric reverse osmosis and nanofiltration membranes. *J. Membr. Sci.* **119**(2), 253–268.

Childress, A. E. and Elimelech, M. (2000) Relating nanofiltration membrane performance to membrane charge (electrokinetic) characteristics. *Environ. Sci. Technol.* **34**(17), 3710–3716.

Cho, J. W., Amy, G. and Pellegfino, J. (1999) Membrane filtration of natural organic matter: initial comparison of rejection and flux decline characteristics with ultrafiltration and nanofiltration membranes. *Water Res.* **33**(11), 2517–2526.

Choi, J.-H., Fukushi, K. and Yamamoto, K. (2008) A study on the removal of organic acids from wastewaters using nanofiltration membranes. *Sep. Purif. Technol.* **59**(1), 17–25.

Choi, J.-H., Jegal, J. and Kim, W.-N. (2006) Fabrication and characterization of multi-walled carbon nanotubes/polymer blend membranes. *J. Membr. Sci.* **284**(1–2), 406–415.

Choi, S., Yun, Z., Hong, S. and Ahn, K. (2001) The effect of co-existing ions and surface characteristics of nanomembranes on the removal of nitrate and fluoride. *Desalination* **133**(1), 53–64.

Chung, T.-S., Qin, J.-J., Huan, A. and Toh, K.-C. (2002) Visualization of the effect of shear rate on the outer surface morphology of ultrafiltration membranes by AFM. *J. Membr. Sci.* **196**(2), 251–266.

Conlon, W. J. and McClellan, S. A. (1989) Membrane softening: treatment process comes of age. *J. AWWA* **81**(11), 47–51.

Dai, Y., Jian, X., Zhang, S. and Guiver, M. D. (2002) Thin film composite (TFC) membranes with improved thermal stability from sulfonated poly(phthalazinone ether sulfone ketone) (SPPESK). *J. Membr. Sci.* **207**(2), 189–197.

De Witte, J. P. (1997) Surface water potabilisation by means of a novel nanofiltration element. *Desalination* **108**(1–3), 153–157.

Desai, T. A., Hansford, D. and Ferrari, M. (1999) Characterization of micromachnined silicon membranes for immunoisolation and bioseparation applications. *J. Membr. Sci.* **159**(1–2), 221–231.

Deshmukh, S. S. and Childress, A. E. (2001) Zeta potential of commercial RO membranes: Influence of source water type and chemistry. *Desalination* **140**(1), 87–95.

Di, Z.-C., Ding, J., Peng, X.-J., Li, Y.-H., Luan, Z.-K. and Liang, J. (2006) Chromium adsorption by aligned carbon nanotubes supported ceria nanoparticles. *Chemosphere* **62**(5), 861–865.

Drioli, E., Criscuoli A. and Curcioa, E. (2002) Integrated membrane operations for seawater desalination. *Desalination* **147**(1–3), 77–81.

Dueom, G. and Cabassud, C. (1999) Interests and limitations of nanofiltration for the removal of volatile organic compounds in drinking water production. *Desalination* **124**(1–3), 115–123.

Duran F. E. and Dunkelberger G. W. (1995) A comparison of membrane softening on three South Florida groundwaters. *Desalination* **102**(1–3), 27–34.

Elimelech, M., Zhu, X., Childress, A. E., and Hong, S. (1997) Role of membrane surface morphology in colloidal fouling of cellulose acetate and composite aromatic polyamide reverse osmosis membranes. *J. Membr. Sci.* **127**(1), 101–109.

El-Sheikh, A. H., Sweileh, J. A. and Al-Degs, Y. S. (2007) Effect of dimensions of multi-walled carbon nanotubes on its enrichment efficiency of metal ions from environmental waters. *Anal. Chim. Acta* **604**(2), 119–126.

Ericsson, B., Hallberg, M. and Wachenfeldt, J. (1996) Nanofiltration of highly colored

raw water for drinking water production. *Desalination* **108**(1–3), 129–141.
Eriksson, P. (1988) Nanofiltration extends the range of membrane filtration. *Environ. Prog.* **7**(1), 58–62.
Escobar, I. C., Hong, S. and Randall, A. (2000) Removal of assimilable and biodegradable dissolved organic carbon by reverse osmosis and nanofiltration membranes. *J. Membr. Sci.* **175**(1), 1–17.
Everest, W. R. and Malloy, S. (2000) A design/build approach to deep aquifer membrane treatment in Southern California. *Desalination* **132**(1–3), 41–45.
Faller, K. A. (1999) Reverse Osmosis and Nanofiltration. *AWWA Manual of Water Supply Practice,* M46.
Flemming, H. C. (1993) Mechanistic aspects of reverse osmosis membrane biofouling and prevention. In Reverse Osmosis: Membrane Technology, Water Chemistry and Industrial Applications (ed. Amjad, Z.), Van Nostrand Reinhold, New York, pp. 163–209.
Freger, V., Arnot, A. C. and Howell, J. A. (2000) Separation of concentrated organic/in organic salt mixtures by nanofiltration. *J. Membr. Sci.* **178**(1–2), 185–193.
Fu, P., Ruiz, H., Lozier, J., Thompson, K. and Spangenberg, C. (1994) Selecting membranes for removing NOM and DBP precursors. *J. AWWA* **86**(12), 55–72.
Fu, P., Ruiz, H., Lozier, J., Thompson, K. and Spangenberg, C. (1995) A pilot study on groundwater natural organics removal by low-pressure membranes. *Desalination* **102**(1–3), 47–56.
Gauden, P. A., Terzyk, A. P., Rychlicki, G., Kowalczyk, P., Lota, K., Raymundo-Pinero, E., Frackowiak, E. and Beguin, F. (2006) Thermodynamic properties of benzene adsorbed in activated carbons and multi-walled carbon nanotubes. *Chem. Phys. Lett.* **421**(4–6), 409–414.
Ghosh, K. and Schnitzer, M. (1980) Macromolecular structures of humic substances. *Soil Sci.* **129**(5), 266–276.
Hafiarle, A., Lemordant, D. and Dhahbi, M. (2000) Removal of hexavalent chromium by nanofiltration. *Desalination* **130**(3), 305–312.
Hagmeyer, G. and Gimbel, R. (1998) Modelling the salt rejection of nanofiltration membranes for ternary ion mixtures and for single salts at different pH values. *Desalination* **117**(1–3), 247–256.
Hamed, O. A. (2005) Overview of hybrid desalination systems – current status and future prospects. *Desalination* **186**(1–3), 207–214.
Hassan, A. M., Al-Sofi, M. A. K., Al-Amoudi, A. S., Jamaluddin, A. T. M., Farooque, A. M., Rowaili, A., Dalvi, A. G. I., Kither, N. M., Mustafa, G. M. and Al-Tisan, I. A. R. (1998) A new approach to thermal seawater desalination processes using nanofiltration membranes (Part 1). *Desalination* **118**(1–3), 35–51.
Hassan, A. M., Farooque, A. M., Jamaluddin, A. T. M., Al-Amoudi, A. S., Al-Sofi, M. A., Al-Rubaian, A. F., Kither, N. M., Al-Tisan, I. A. R. and Rowaili, A. (2000) A demonstration plant based on the new NF-SWRO process. *Desalination* **131**(1–3), 157–171.
Hirose, M., Ito, H. and Kamiyama, Y. (1996) Effect of skin layer surface structures on the flux behavior of RO membrane. *J. Membr. Sci.* **121**(2), 209–215.
Huiting, H., Kappelhof, J. W. N. M. and Bosklopper, Th. G. J. (2001) Operation of NF/RO plants: from reactive to proactive. *Desalination* **139**(1–3), 183–189.
Hvid, K. B., Nielsen, P. S. and Stengaard, F. F. (1990) Preparation and characterization of a new ultrafiltration membrane. *J. Membr. Sci.* **53**(3), 189–202.
Iijima, S. (1991) Helical microtubules of graphitic carbon. *Nature* **354**, 56–58.
Jacob, M., Guigui, C., Cabassud, C., Darras, H., Lavison, G. and Moulin, L. (2009) Performances of RO and NF processes for wastewater reuse: Tertiary treatment after

a conventional activated sludge or a membrane bioreactor. *Desalination* **250**(2), 833–839.

Jarusutthirak, C., Amy G. and Croué, J.-P. (2002) Fouling characteristics of wastewater effluent organic matter (EfOM) isolates on NF and UF membranes. *Desalination* **145**(1–3), 247–255.

Jegal, J., Min, S. G. and Lee, K.-H. (2002) Factors affecting the interfacial polymerization of polyamide active layers for the formation of polyamide composite membranes. *J. Appl. Polym. Sci.* **86**(11), 2781–2787.

Jian, X., Dai, Y., He, G. and Chen, G. (1999) Preparation of UF and NF poly(phthalazine ether sulfone ketone) membranes for high temperature application. *J. Membr. Sci.* **161**(1–2), 185–191.

Jönsson, A. and Jönsson, B. (1991) The influence of nonionic surfactants on hydrophobic and ultrafiltration membranes. *J. Membr. Sci.* **56**(1), 49–76.

Jung, Y. J., Kiso, Y., Othman, R. A. A. B., Ikeda, A., Nishimura, K., Min, K. S., Kumano, A. and Ariji, A. (2005) Rejection properties of aromatic pesticides with a hollow-fiber NF membrane. *Desalination* **180**(1–3), 63–71.

Karime, M., Bouguecha, S. and Hamrouni, B. (2008) RO membrane autopsy of Zarzis brackish water desalination plant. *Desalination* **220**(1–3), 258–266.

Kartinen, E. O. and Martin, C. J. (1995) An overview of arsenic removal processes. *Desalination* **103**(1–2), 79–88.

Keszler, B., Kovács, G., Tóth, A., Bertóti, I. and Hegyi, M. (1991) Modified polyethersulfone membranes. *J. Membr. Sci.* **62**(2), 201– 210.

Khalik, A. and Praptowidodo, V. S. (2000) Nanofiltration for drinking water production from deep well water. *Desalination* **132**(1–3), 287–292.

Khirani, S., Ben Aim, R. and Manero, M.-H. (2006) Improving the measurement of the Modified Fouling Index using nanofiltration membranes (NF–MFI). *Desalination* **191**(1–3), 1–7.

Kim, I. C., Lee, K.-H. and Tak, T.-M. (2001) Preparation and characterization of integrally skinned uncharged polyetherimide asymmetric nanofiltration membrane. *J. Membr. Sci.* **183**(2), 235–247.

Kim, I.-C., Jegal, J. and Lee, K.-H. (2002) Effect of aqueous and organic solutions on the performance of polyamide thin-film-composite nanofiltration membranes. *J. Polym. Sci.: Part B* **40**(19), 2151–2163.

Kim, K. J., Fane, A. G. and Fell, C. J. D. (1988) The performance of ultrafiltration membranes pretreated by polymers. *Desalination* **70**(1–3), 229–249.

Kim, K. J., Fane, A. G. and Fell, C. J. D. (1989) The effect of Langmuir Blodgett layer pretreatment on the performance of ultrafiltration membranes. *J. Membr. Sci.* **43**(2–3), 187–204.

Kim, M., Saito, K. and Furusaki, S. (1991) Water flux and protein adsorption of a hollow fiber modified with hydroxyl groups. *J. Membr. Sci.* **56**(3), 289–302.

Kim, Y.-H., Hwang, E.-D., Shin, W. S., Choi, J.-H., Ha, T. W. and Choi, S. J. (2007) Treatments of stainless steel wastewater containing a high concentration of nitrate using reverse osmosis and nanomembranes. *Desalination* **202**(1–3), 286–292.

Kimura, K., Amy, G., Drewes, J. and Watanabe, Y. (2003b) Adsorption of hydrophobic compounds onto NF/RO membranes: An artifact leading to overestimation of rejection. *J. Membr. Sci.* **221**(1–2), 89–101.

Kimura, K., Amy, G., Drewes, J., Heberer, T., Kim, T.-U. and Watanabe, Y. (2003a) Rejection of organic micropollutants (disinfection by-products, endocrine disrupting compounds, and pharmaceutically active compounds) by NF/RO membranes. *J. Membr. Sci.* **227**(1–2), 113–121.

Kiso, Y., Kitao, T., Kiyokatsu, J. and Miyagi M. (1992) The effects of molecular width on permeation of organic solute through cellulose acetate reverse osmosis membrane. *J. Membr. Sci.* **74**(1–2), 95–103.

Kiso, Y., Kon, T., Kitao, T. and Nishimura, K. (2001a) Rejection properties of alkyl phthalates with nanofiltration membranes. *J. Membr. Sci.* **182**(1–2), 205–214.

Kiso, Y., Nishimura, Y., Kitao, T. and Nishimura, K. (2000) Rejection properties of non-phenylic pesticides with nanofiltration membranes. *J. Membr. Sci.* **171**(2), 229–237.

Kiso, Y., Sugiura, Y., Kitao, T. and Nishimura, K. (2001b) Effects of hydrophobicity and molecular size on rejection of aromatic pesticides with nanofiltration membranes. *J. Membr. Sci.* **192**(1–2), 1–10.

Košutić, K. and Kunst, B. (2002) Removal of organics from aqueous solutions by commercial RO and NF membranes of characterized porosities. *Desalination* **142**(1), 47–56.

Košutić, K., Furač, L., Sipos, L. and Kunst, B. (2005) Removal of arsenic and pesticides from drinking water by nanofiltration membranes. *Sep. Purif. Technol.* **42**(2), 137–144.

Košutić, K., Kaštelan-Kunst, L. and Kunst, B. (2000) Porosity of some commercial reverse osmosis and nanofiltration polyamide thin-film composite membranes. *J Membr Sci* **168**(1–2), 101–108.

Ku, Y., Chen, S.-W. and Wang, W.-Y. (2005) Effect of solution composition on the removal of copper ions by nanofiltration. *Sep. Purif. Technol.* **43**(2), 135–142.

Kumakiri, I., Yamaguchi, T. and Nakao, S. (2000) Application of a zeolite A membrane to reverse osmosis process. *J. Chem. Eng. Jpn.* **33**, 333.

Larbot, A., Alami-Younssi, S., Persin, M., Sarrazin, J. and Cot, L. (1994) Preparation of a γ-alumina nanofiltration membrane. *J. Membr. Sci.* **97**, 167–173.

Le Roux, I., Krieg, H. M., Yeates, C. A. and Breytenbach, J. C. (2005) Use of chitosan as an antifouling agent in a membrane bioreactor. *J. Membr. Sci.* **248**(1–2), 127–136.

Lee, S., Park, G., Amy, G., Hong, S.-K., Moon, S.-H., Lee, D.-H. and Cho, J. (2002) Determination of membrane pore size distribution using the fractional rejection of nonionic and charged macromolecules. *J. Membr. Sci.* **201**(1–2), 191–201.

Lee, S., Quyet, N., Lee, E., Kim, S., Lee, S., Jung, Y. D., Choi, S. H. and Cho, J. (2008) Efficient removals of tris(2-chloroethyl) phosphate (TCEP) and perchlorate using NF membrane filtrations. *Desalination* **221**(1–3), 234–237.

Levine, B. B., Madireddi, K., Lazarova, V., Stenstrom, M. K. and Suffet, M. (1999) Treatment of trace organic compounds by membrane processes: At the lake arrowhead water reuse pilot plant. *Water. Sci. Technol.* **40**(4–5), 293–301.

Li, L., Dong, J., Nenoff, T. M. and Lee, R. (2004a) Desalination by reverse osmosis using MFI zeolite membranes. *J. Membr. Sci.* **243**(1–2), 401–404.

Li, L., Dong, J., Neoff, T. M. and Lee, R. (2004b) Reverse osmosis of ionic aqueous solutions on a MFI zeolite membrane. *Desalination* **170**(3), 309–316.

Li, Y.-H., Wang, S., Wei, J., Zhang, X., Xu, C., Luan, Z., Wu, D. and Wei, B. (2002) Lead adsorption on carbon nanotubes. *Chem. Phys. Lett.* **357**(3–4), 263–266.

Li, Y.-H., Wang, S., Zhang, X., Wei, J., Xu, C., Luan, Z. and Wu, D. (2003) Adsorption of fluoride from water by aligned carbon nanotubes. *Mater. Res. Bull.* **38**(3), 469–476.

Lin, J. and Murad, S. (2001a) The role of external electric fields in membrane-based separation processes: A molecular dynamics study. *Mol. Phys.* **99**(5), 463–469.

Lin, J. and Murad, S. (2001b) A computer simulation study of the separation of aqueous solution using thin zeolite membranes. *Mol. Phys.* **99**(14), 1175–1181.

Long, R. Q. and Yang, R. T. (2001) Carbon nanotubes as superior sorbent for dioxin removal. *J. Am. Chem. Soc.* **123**(9), 2058–2059.

Lu, C., Chung, Y.-L. And Chang, K.-F. (2005) Adsorption of trihalomethanes from water with carbon nanotubes. *Water Res.* **39**(6), 1183–1189.

Lu, X., Bian, X. and Shi, L. (2002) Preparation and characterization of NF composite membrane. *J. Membr. Sci.* **210**(1), 3–11.

Luyten, J., Cooymans, J., Smolders, C., Vercauteren, S., Vansant, E. F. and R. Leysen (1997) Shaping of multilayer ceramic membranes by dip-coating. *J. Eur. Ceram. Soc.* **17**(2–3), 273–279.

Madireddi, K., Babcock, R. W., Levine, B., Huo, T. L., Khan, E., Ye, Q. F., Neethling, J. B., Suffet, I. H. and Stenstrom, M. K. (1997) Wastewater reclamation at Lake Arrowhead, California: An overview. *Water Environ. Res.* **69**(3), 350–362.

Majewska-Nowak, K., Kabsch-Korbutowicz, M., Dodź M. and Winnicki, T. (2002) The influence of organic carbon concentration on atrazine removal by UF membranes. *Desalination* **147**(1–3), 117–122.

Martin, F., Walczak, R., Boiarski, A., Cohen, M., West, T., Cosentino, C. and Ferrari, M. (2005) Tailoring width of microfabricated nanochannels to solute size can be used to control diffusion kinetics. *J. Control. Release* **102**(1), 123–133.

Mänttäri, M., Puro, L., Nuortila-Jokinen, J. and Nyström, M. (2000) Fouling effects of polysaccharides and humic acid in nanofiltration. *J. Membr. Sci.* **165**(1), 1–17.

Mohammad, A. W. and Ali, N. (2002) Understanding the steric and charge contributions in NF membranes using increasing MWCO polyamide membranes. *Desalination* **147**(1–3), 205–212.

Mohesn, M. S., Jaber J. O. and Afonso, M. D. (2003) Desalination of brackish water by nanofiltmtion and reverse osmosis. *Desalination* **157**(1–3), 167.

Montovay, T., Assenmacher, M. and Frimmel, F. H. (1996) Elimination of pesticides from aqueous solution by nanofiltration. *Magyar Kémiai Folyóirat* **102**(5), 241–247.

Moon, E. J., Seo, Y. S. and Kim, C. K. (2004) Novel composite membranes prepared from 2,2 bis [4-(2-hydroxy-3-methacryloyloxy propoxy) phenyl] propane, triethylene glycol dimethacrylate, and their mixtures for the reverse osmosis process. *J. Membr. Sci.* **243**(1–2), 311–316.

Murad, S. and Nitche, L. C. (2004) The effect of thickness, pore size and structure of a nanomembrane on the flux and selectivity in reverse osmosis separations: A molecular dynamics study. *Chem. Phys. Lett.* **397**(1–3), 211–215.

Murad, S., Jia, W. and Krishnamurthy, M. (2004) Ion-exchange of monovalent and bivalent cations with NaA zeolite membranes: a molecular dynamics study. *Mol. Phys.* **102**(19), 2103–2112.

Murad, S., Oder, K. and Lin, J. (1998) Molecular simulation of osmosis, reverse osmosis, and electro-osmosis in aqueous and methanolic electrolyte solutions. *Mol. Phys.* **95**(3), 401–408.

Nederlof, M. M., Kruithof, J. C., Taylor, J. S., Van Der Kooij, D. and Schippers, J. C. (2000) Comparison of NF/RO membrane performance in integrated membrane systems. *Desalination* **131**(1–3), 257–269.

Nghiem, L. D., Schäfer, A. I. and Elimilech M. (2004) Removal of natural hormones by nanofiltration membranes: Measurement, modeling, and mechanisms. *Environ. Sci. Technol.* **38**(6), 1888–1896.

Nghiem, L. D., Schäfer, A. I. and Waite, T. D. (2002) Adsorption of estrone on nanofiltration and reverse osmosis membranes in water and wastewater treatment.

Water Sci. Technol. **46**(4–5), 265–272.
Nunes, S. R. and Peinemann, K. V. (2001) In *Membrane Technology in the Chemical Industry*, 1st edn (eds. Nunes, S. R. and Peinemann, K. V.), Wiley-VCH, Germany, pp. 1–53.
Nystrom, M. and Jarvinen, P. (1991) Modification of polysulfone ultrafiltration membranes with UV irradiation and hydrophilicity increasing agents. *J. Membr. Sci.* **60**(2–3), 275–296.
Ortega, L. M., Lebrun, R., Blais, J.-F. and Hausler, R. (2008) Removal of metal ions from an acidic leachate solution by nanofiltration membranes. *Desalination* **227**(1–3), 204–216.
Otaki, M., Yano, K. and Ohgaki, S. (1998) Virus removal in a membrane separation process. *Water Sci. Technol.* **37**(10), 107–116.
Ozaki, H. and Li, H. (2002) Rejection of organic compounds by ultralow pressure reverse osmosis membrane. *Water Res.* **36**(1), 123–130.
Paulose, M., Peng, L., Popat, K. C., Varghese, O. K., LaTempa, T. L., Bao, N., Desai, T. A. and Grimes, C. A. (2008) Fabrication of mechanically robust, large area, polycrystalline nanotubular/ porous TiO_2 membranes. *J. Membr. Sci.* **319**(1–2), 199–205.
Peng, X., Li, Y., Luan, Z., Di, Z., Wang, H., Tian, B. and Jia, Z. (2003) Adsorption of 1,2-dichlorobenzene from water to carbon nanotubes. *Chem. Phys. Lett.* **376**(1–2), 154–158.
Petersen, R. J. (1993) Composite reverse osmosis and nanofiltration membranes. *J. Membr. Sci.* **83**(1), 81–150.
Plakas, K. V., Karabelas, A. J., Wintgens, T. and Melin, T. (2006) A study of selected herbicides retention by nanofiltration membranes – The role of organic fouling. *J. Membr. Sci.* **284**(1–2), 291–300.
Pontié, M., Diawara, C., Rumeau, M., Aurean D. and Hemmerey, P. (2003) Seawater nanofiltration (NF): Fiction or reality? *Desalination* **158**(1–3), 277–280.
Rabie, H. R., Côté P. and Adams, N. (2001) A method for assessing membrane fouling in pilot- and full-scale systems. *Desalination* **141**(3), 237–243.
Rao, A. P., Desai, N. V. and Rangarajan, R. (1997) Interfacially synthesized thin film composite RO membranes for seawater desalination. *J. Membr. Sci.* **124**(2), 263–272.
Redondo, J. A. (2001) Lanzarote IV, a new concept for two-pass SWRO at low O&M cost using the new high-flow FILMTEC SW30-380. *Desalination* **138**(1–3), 231–236.
Reiss, C. R., Taylor, J. S. and Robert, C. (1999) Surface water treatment using nanofiltration – pilot testing results and design considerations. *Desalination* **125**(1–3), 97–112.
Ridgway, H. F. and Flemming, H. F. (1996) Membrane biofouling. In *Water Treatment Membrane Processes* (eds. Mallevialle, J. Odendaal, P. E. and Wiesner, M. R.), McGraw-Hill, New York, pp. 6.1–6.62.
Roh, I. J., Park, S. Y., Kim, J. J. and Kim, C. K. (1998) Effects of the polyamide molecular structure on the performance of reverse osmosis membranes. *J. Polym. Sci.: Part B* **36**(11), 1821–1830.
Roorda J. H. and Van der Graaf, J. H. J. M. (2001) New parameter for monitoring fouling during ultrafiltration of WWTP effluent. *Water Sci. Technol.* **43**(10), 241–248.
Roudman, A. R. and DiGiano, F. A. (2000) Surface energy of experimental and commercial nanofiltration membranes: Effects of wetting and natural organic matter fouling. *J. Membr. Sci.* **175**(1), 61–73.

Saitúa, H., Campderrós, M., Cerutti S. and Pérez Padilla, A. (2005) Effect of operating conditions in removal of arsenic from water by nanofiltration membrane. *Desalination* **172**(2), 173–180.

Sato, Y., Kang, M., Kamei T. and Magara, Y. (2002) Performance of nanofiltration for arsenic removal. *Water Res.* **36**(13), 3371–3377.

Schäfer, A. I., Fane, A. G. and Waite, T. (1998) Nanofiltration of natural organic matter: Removal, fouling and the influence of multivalent ions. *Desalination* **118**(1–3), 109–122.

Schäfer, A. I., Mastrup, M. and LundJensen, R. (2002) Particle interactions and removal of trace contaminants from water and wastewaters. *Desalination* **147**(1–3), 243–250.

Schäfer, A. I., Pihlajamäki, A., Fane, A. G., Waite, T. D. and Nyström M. (2004) Natural organic matter removal by nanofiltration: Effects of solution chemistry on retention of low molar mass acids versus bulk organic matter. *J. Membr. Sci.* **242**(1–2), 73–85.

Schippers, J. C. and Verdouw, J. (1980) The modified fouling index, a method of determining the fouling characteristics of water. *Desaleation* **32**, 137–148.

Schneider, R. P., Ferreira, L. M., Binder, P., Bejarano, E. M., Góes, K. P., Slongo, E., Machado, C. R. and Rosa, G. M. Z. (2005) Dynamics of organic carbon and of bacterial populations in a conventional pretreatment train of a reverse osmosis unit experiencing severe biofouling. *J. Membr. Sci.* **266**(1–2), 18–29.

Schutte, C. F. (2003) The rejection of specific organic compounds by reverse osmosis membranes. *Desalination* **158**(1–3), 285–294.

Seidel, A., Waypa, J. J. and Elimech, M. (2001) Role of charge (Donnan) exclusion in removal of arsenic from water by a negatively charged porous nanofiltration membrane. *Environ. Eng. Sci.* **18**(2), 105–113.

Semiat, R. (2000) Desalination: Present and future. *Water Internet.* **25**(1), 54–65.

Shih, M. C. (2005) An overview of arsenic removal by pressure-driven membrane processes. *Desalination* **172**(1), 85–97.

Shim, Y., Lee, H.-G., Lee, S., Moon, S.-H. and Cho, J. (2002) Effects of natural organic matter and ionic species on membrane surface charge. *Environ. Sci. Technol.* **36**(17), 3864–3871.

Sikora, J., Hansson C. H. and Ericsson, B. (1989) Pre-treatment and desalination of mine drainage water in a pilot plant. *Desalination* **75**, 363–373.

Simon, A., Nghiem, L. D., Le-Clech, P., Khan, S. J. and Drewes, J. E. (2009) Effects of membrane degradation on the removal of pharmaceutically active compounds (PhACs) by NF/RO filtration processes. *J. Membr. Sci.* **340**(1–2), 16–25.

Sombekke, H. D. M., Voorhoeve, D. K. and Hiemstra, P. (1997) Environmental impact assessment of groundwater treatment with nanofiltration. *Desalination* **113**(2–3), 293–296.

Song, Y. J., Liu, F. and Sun, B. H. (2005) Preparation, characterization, and application of thin film composite nanofiltration membranes. *J. Appl. Polym. Sci.* **95**(3), 1251–1261.

Soria, R. and Cominotti, S. (1996) Nanofiltration ceramic membrane. In *Proc. of the International Conference on the Membranes and Membrane Processes,* Yokohama, Japan, 18–23 August.

Srivastava, A., Srivastava, O. N., Talapatra, S., Vajtai, R. and Ajayan, P. M. (2004) Carbon nanotube filters. *Nat. Mater.* **3**, 610–614.

Stengaard, F. F. (1988) Characteristics and performance of new types of ultrafiltration membranes with chemically modified surfaces. *Desalination* **70**(1–3), 207–224.

Tang, A. and Chen, V. (2002) Nanofiltration of textile wastewater for water reuse. *Desalination* **143**(1), 11–20.

Tang, C., Zhang, Q., Wang, K., Fu, Q. and Zhang, C. (2009) Water transport behavior of chitosan porous membranes containing multi-walled carbon nanotubes (MWNTs). *J. Membr. Sci.* **337**(1–2), 240–247.

Tanninen, J. and Nystrom, M. (2002) Separation of ions in acidic conditions using NF. *Desalination* **147**(1–3), 295–299.

Tasaka, K., Katsura, T., Iwahori, H. and Kamiyama, Y. (1994) Analysis of RO elements operated at more than 80 plants in Japan. *Desalination* **96**(1–3), 259–272.

Tay, J.-H., Liu, J. and Sun, D. D. (2002) Effect of solution physico-chemistry on the charge property of nanofiltration membranes. *Water Res.* **36**(3), 585–598.

Terrones, M. (2004) Carbon nanotubes: Synthesis and properties, electronic devices and other emerging applications. *Int. Mater. Rev.* **49**(6), 325–377.

Tödtheide, V., Laufenberg, G. and Kunz, B. (1997) Waste water treatment using reverse osmosis: Real osmotic pressure and chemical functionality as influencing parameters on the retention of carboxylic acids in multi-component systems. *Desalination* **110**(3), 213–222.

Ulbricht, M., Matuschewski, H., Oechel, A. and Hicke, H. G. (1996) Photo-induced graft polymerization surface modification for the preparation of hydrophilic and low-protein-adsorbing ultrafiltration membranes. *J. Membr. Sci.* **115**(1), 31–47.

Urase, T., Oh, J. and Yamamoto, K. (1998) Effect of pH on rejection of different species of arsenic by nanofiltration. *Desalination* **117**(1–3), 11–18.

Urase, T., Yamamoto, K. and Ohgaki, S. (1996) Effect of pore structure of membranes and module configuration on virus retention. *J. Membr. Sci.* **115**(1), 21–29.

Van der Bruggen, B. and Vandecasteele, C. (2001b) Flux decline during nanofiltration of organic components in aqueous solution. *Environ. Sci. Technol.* **35**(17), 3535–3540.

Van der Bruggen, B. and Vandecasteele, C. (2002) Modeling of the retention of uncharged molecules with nanofiltration. *Water Res.* **36**(5), 1360–1368.

Van der Bruggen, B. and Vandecasteele, C. (2003) Removal of pollutants from surface water andgroundwater by nanofiltration: Overview of possible applications in the drinking water industry *Environ. Poll.* **122**(3), 435–445.

Van der Bruggen, B., Braeken, L. and Vandecasteele, C. (2002a) Evaluation of parameters describing flux decline in nanofiltration of aqueous solutions containing organic compounds. *Desalination* **147**(1–3), 281–288.

Van der Bruggen, B., Braeken, L. and Vandecasteele, C. (2002b) Flux decline in nanofiltration due to adsorption of organic compounds. *Sep. Purif. Technol.* **29**(1), 23–31.

Van der Bruggen, B., Everaert, K., Wilms, D. and Vandecasteele, C. (2001a) Application of nanofiltration for removal of pesticides, nitrate and hardness from groundwater: Rejection properties and economic evaluation. *J. Membr. Sci.* **193**(2), 239–248.

Van der Bruggen, B., Mänttäri, M. and Nyström, M. (2008) Drawbacks of applying nanofiltration and how to avoid them: A review. *Sep. Purif. Technol.* **63**(2), 251–263.

Van der Bruggen, B., Schaep, J., Maes, W., Wilms, D. and Vandecasteele, C. (1998) Nanofiltration as a treatment method for the removal of pesticides from ground waters. *Desalination* **117**(1–3), 139–147.

Van der Bruggen, B., Schaep J., Wilms D. and Vandecasteele C. (1999) Influence of molecular size, polarity and charge on the retention of organic molecules by nanofiltration. *J. Membr. Sci.* **156**(1), 29–41.

Van Gestel, T., Vandecasteele, C., Buekenhoudt, A., Dotremont, C., Luyten, J., Leysen, R., Van der Bruggen, B. and Maes, G. (2002a) Alumina and titania multilayer membranes for nanofiltration: Preparation, characterization and chemical stability. *J. Membr. Sci.* **207**(1), 73–89.

Van Gestel, T., Vandecasteele, C., Buekenhoudt, A., Dotremont, C., Luyten, J., Leysen, R., Van der Bruggen, B. and Maes, G. (2002b) Salt retention in nanofiltration with multilayer ceramic TiO_2 membranes. *J. Membr. Sci.* **209**(2), 379–389.

Van Hoop, S. C. J. M., Minnery, J. G. and Mack, B. (2001) Dead-end ultrafiltration as alternative pre-treatment to reverse osmosis in seawater desalination: A case study. *Desalination* **139**(1–3), 161–168.

Verissimo, S., Peinemann, K.-V. and Bordado, J. (2005) Thin-film composite hollow fiber membranes: An optimized manufacturing method. *J. Membr. Sci.* **264**(1–2), 48–55.

Verliefde, A., Cornelissen, E., Amy, G., Van der Bruggen, B. and Van Dijk, H. (2007) Priority organic micropollutants in water sources in Flanders and the Netherlands and assessment of removal possibilities with nanofiltration. *Environ. Pollut.* **146**(1), 281–289.

Visvanathan, C., Marsono B. and Basu, B. (1998) Removal of THMP by nanofiltration: Effects of interference parameters. *Water Res.* **32**(12), 3527–3538.

Voigt, I., Stahn, M., Wöhner, St., Junghans, A., Rost, J. and Voigt, W. (2001) Integrated cleaning of coloured wastewater by ceramic NF membranes. *Sep. Purifi. Technol.* **25**(1–3), 509–512.

Vrouwenvelder, J. S., Kappelhof, J. W. N. M., Heijman, S. G. J., Schippers J. C. and Van der Kooija, D. (2003) Tools for fouling diagnosis of NF and RO membranes and assessment of the fouling potential of feed water. *Desalination* **157**(1–3), 361–365.

Wang, X. F., Chen, X. M., Yoon, K., Fang, D. F., Hsiao, B. S. and Chu, B. (2005) High flux filtration medium based on nanofibrous substrate with hydrophilic nanocomposite coating. *Environ. Sci. Technol.* **39**(19), 7684–7691.

Wang, X.-L., Tsuru, T., Nakao, S.-I. and Kimura, S. (1997) The electrostatic and steric-hindrance model for the transport of charged solutes through nanofiltration membranes. *J. Membr. Sci.* **135**(1), 19–32.

Wang, X.-L., Wang, W.-N. and Wang, D.-X. (2002) Experimental investigation on separation performance of nanofiltration membranes for inorganic electrolyte solutions. *Desalination* **145**(1–3), 115–122.

Waniek, A., Bodzek, M. and Konieczny, K. (2002) Trihalomethanes removal from water using membrane processes. *Polish J. Environ. Studies* **11**(2), 171–178.

Waypa, J. J., Elimelech, M. and Hering, J. G. (1997) Arsenic removal by RO and NF membranes. *J. AWWA* **89**(10), 102–114.

Wintgens, T., Gallenkemper, M. and Melin, T. (2003) Occurrence and removal of endocrine disrupters in landfill leachate treatment plants. *Water Sci. Technol.* **48**(3), 127–134.

Wittmann, E., Cote, P., Medici, C., Leech, J. and Turner, A. G. (1998) Treatment of a hard borehole water containing low levels of pesticide by nanofiltration. *Desalination* **119**(1–3), 347–352.

Wu, C., Zhang, S., Yang, D. and Jian, X. (2009) Preparation, characterization and application of a novel thermal stable composite nanofiltration membrane. *J. Membr. Sci.* **326**(2), 429–434.

Xia, S. J., Dong, B. Z., Zhang, Q. L., Xu, B., Gao, N. Y. and Causserand, C. (2007) Study of arsenic removal by nanofiltration and its application in China. *Desalination* **204**(1–3), 374–379.

Xu, P., Drewes, J. E., Bellona, C., Amy, G., Kim, T.-U., Adam, M., and Heberer, T. (2005) Rejection of emerging organic micropollutants in nanofiltration-reverse osmosis membrane applications. *Water Environ. Res.* **77**(1), 40–48.

Xu, P., Drewes, J. E., Kim, T.-U., Bellona, C. and Amy, G. (2006) Effect of membrane fouling on transport of organic contaminants in NF/RO membrane applications. *J Membr. Sci.* **279**(1–2), 165–175.

Xu, Y. and Lebrun, R. E. (1999) Investigation of the solute separation by charged nanofiltration membrane: Effect of pH, ionic strength and solute type. *J Membr. Sci.* **158**(1–2), 93–104.

Yahya, M. T., Blu, C. B. and Gerha, C. P. (1993) Virus removal by slow sand filtration and nanofiltration. *Water Sci. Technol.* **27**(3–4), 445–448.

Yamagishi, H., Crivello, J. V. and Belfort, G. (1995) Development of a novel photochemical technique for modifying poly(arysulfone) ultrafiltration membranes. *J. Membr. Sci.* **105**(3), 237–247.

Yang, K., Zhu, L. and Xing, B. (2006) Adsorption of polycyclic aromatic hydrocarbons by carbon nanomaterials, *Environ. Sci. Technol.* **40**(6), 1855–1861.

Yeh, H.-H., Tseng, I-C., Kao, S.-J., Lai, W.-L., Chen, J.-J., Wang, G. T. and Lin, S.-H. (2000) Comparison of the finished water quality among an integrated membrane process, conventional and other advanced treatment processes. *Desalination* **131**(1–3), 237–244.

Ying, L., Zhai, G., Winata, A. Y., Kang, E. T. and Neoh, K. G. (2003) pH effect of coagulation bath on the characteristics of poly(acrylic acid)-grafted and poly(4-vinylpyridine)-grafted poly(vinylidene fluoride) microfiltration membranes. *J. Colloid Interface Sci.* **265**(2), 396–403.

Yoon, S.-H., Lee, C.-H., Kim, K.-J. and Fane, A. G. (1998) Effect of calcium ion on the fouling of nanofilter by humic acid in drinking water production. *Water Res.* **32**(7), 2180–2186.

Yoon, Y., Amy, G., Cho, J., Her, N. and Pellegrino, J. (2002) Transport of perchlorate (ClO_4^-) through NF and UF membranes. *Desalination* **147**(1–3), 11–17.

Yoon, Y., Westerhoff, P., Snyder, S. A., Wert, E. C. and Yoon, J. (2007) Removal of endocrine disrupting compounds and pharmaceuticals by nanofiltration and ultrafiltration membranes. *Desalination* **202**(1–3), 16–23.

Zhang, H., Quan, X., Chen, S., Zhao, H. and Zhao, Y. (2006) The removal of sodium dodecylbenzene sulfonate surfactant from water using silica/titania nanorods/nanotubes composite membrane with photocatalytic capability. *Appl. Surf. Sci.* **252**(24), 8598–8604.

Zhang, X., Du, A. J., Lee, P., Sun, D. D. and Leckie, J. O. (2008) Grafted multifunctional titanium dioxide nanotube membrane: Separation and photodegradation of aquatic pollutant. *Appl. Catal. B* **84**(1–2), 262–267.

Zhang, Y., Causserand, C., Aimar, P. and Cravedi, J. P. (2006) Removal of bisphenol A by a nanofiltration membrane in view of drinking water production. *Water Res.* **40**(20), 3793–3799.

Zhao, Y., Taylor, J. and Hong, S. (2005) Combined influence of membrane surface properties and feed water qualities on RO/NF mass transfer, a pilot study. *Water Res.* **39**(7), 1233–1244.

5

Ion Exchange and Adsorption Techniques for Micropollutant Treatment

N. Chubar

5.1 INTRODUCTION

Adsorptive phenomena are widely spread throughout nature. Mountain rocks and soils are just huge columns filled with adsorbents through which the water and gaseous solution flow. Lung tissue behaves very similarly to an adsorbent; it is a carrier for blood hemoglobin providing the transfer of oxygen to the organism. Many functions of the biological membranes of living cells are linked to their surface properties; for instance: the total square of biologically active membranes in organisms reaches a few thousand square metres. Even the senses of taste and smell of human beings depend on adsorption of the related molecules by the surface of the nose and tongue.

The complexity of the adsorptive phenomena and a variety of their applications in nature, society, sciences, and technology are reflected in slightly different definitions of adsorption which point at the process, its nature, or its application. *Sci-Tech Encyclopedia* writes that "adsorption is a process in which atoms or molecules move from a bulk phase (that is, solid, liquid, or gas) onto a solid or liquid surface". *Dental dictionary* explains that "adsorption is a natural process whereby molecules of a gas or liquid adhere to the surface of a solid". The *Britannica Concise Encyclopedia* defines adsorption as a capability of a solid substance (adsorbent) to attract to its surface molecules of a gas or solution (adsorbate) with which it is in contact. *Architecture dictionary* explains adsorption as "the action of a material in extracting a substance from the atmosphere (or a mixture of gases and liquids) and gathering it on the surface in a condensed layer". *Geography dictionary* gives a more practical: "... in soil science, Adsorption is the addition of ions or molecules to the electrically charged surface of a particle of clay or humus". *Veterinary dictionary* defines briefly that "adsorption is the action of a substance in attracting and holding other materials or particles on its surface". And the most practical definition of adsorption is given by business oriented website (TeachMeFinance.com) showing mostly the technological side of adsorption which is "[the] removal of a pollutant from air or water by collecting the pollutant on the surface of a solid material; e.g., an advanced method of treating waste in which activated carbon removes organic matter from waste-water".

The list of definitions is much longer but the few mentioned above are sufficient to conclude that adsorption is complex process involving different reactions (mechanisms); it is widespread in the environment, human society (medicine), and technological processes.

Going a little bit deeper on the scientific level we have to add that adsorption is a concentrating of gaseous or dissolved substances at the interface between two phases: on the surface of a solid or liquid. It is one of the surface phenomena taking place at the interface due to an excess of free energy. Adsorption is a spontaneous process which is accompanied by a decrease of Gibbs energy (at $p =$ const.) or a decrease of Helmholtz energy (at $V =$ const.). However, there is no equalization of concentrations in the whole volume. On the contrary, the difference between the concentrations in the gas/liquid and solid phases increases. At the same time, mobility of adsorbed material to the surface molecules reduces. Both of these factors lead to a decrease of entropy ($\Delta S < 0$).

At present, adsorption is the basis of many industrial operations and scientific investigations. One of the most important fields of research and application of adsorptive processes is the purification, concentrating, and separation of different substances, as well as, adsorptive gaseous and liquid chromatography.

This process is an important stage of heterogeneous catalysis and corrosion. Surface investigations are linked closely to the development of semi-conductor technology, medicine, construction, and military affairs. Also, Aadsorptive processes play key roles when strategic techniques of environmental protection must be chosen.

5.2 THE MAIN STAGES OF ADSORPTION & ION EXCHANGE SCIENCE DEVELOPMENT

The origin of Adsorption Science is dated to the second part of the 18th century which was also a break-though time in the development of chemistry and a transition period from Natural Philosophy and Alchemy (which were looking for universal catalysts (e.g., philosophy stone) as well as a universal solvent (able to transfer any metal into silver and gold) to modern Chemistry and Physical Chemistry. Adsorption science started from the studies of the three greatest chemists: F. Fontana, K.W. Sheele and J.T. Lovitz.

In 1777 looking for phlogiston (Conant, 1950) – like most chemists of that time – Italian chemist, Professor of Pisa University, Felice Fontana (who is also defined as the greatest Italian chemist of that century (Id. Science, 1991)) and Swedish chemist, Karl Wilhelm Scheele (Shectman, 2003) discovered independently the ability of wood coal to take up gases. Later, in 1785, academician Johann Tobias Lovitz (a naturalized German who had lived in St. Petersburg (Russia) since his childhood) discovered the phenomenon of adsorption by charcoal from water solutions (Zolotov, 1998). First, J.T. Lovitz noticed the ability of such coals to discolor the water solutions of organic acids, however, being a real scientist, he did not stop his study there and carried out many additional experiments investigating charcoal as an adsorbing agent. First of all, he tested carbon powders for purification of different contaminated water solutions. He found out that carbons purified a variety of duty (e.g., brown) solutions, made the color of different juices and honey much lighter and discolored the solution of painting substances. J.T. Lovitz studied the action of charcoal on the solution with some odors and discovered that charcoal purified spoilt, smelly water, making it suitable for drinking. He tested the carbon powder for garlic and even for bugs noticing that the charcoal reduced their unpleasant smell. J.T. Lovitz realized the practical applications quickly. In 1790 he published an article entitled: "Guide on the new method how to make a spoilt drinking water suitable for drinking during sea routes". His method was applied by the Russian navy.

The phenomenon of ion exchange was discovered around 150 years ago. In 1845 the English soil scientist H.M. Thompson was passing an ammonia-containing solution through some ordinary garden soil and discovered that the ammonia content of the liquid manure was greatly reduced (Thompson, 1850; Lucy, 2003). It was shown later that the reason for the ion exchange properties of soil was the presence of fine particles of a natural material called zeolite which had ion exchange properties. Since that time, the interest in ion exchange has not been weakened but has been growing permanently. The significance of these processes in agricultural chemistry and the chemistry of living (including human) organisms has been increasingly realized. The principles of ion exchange have been extensively applyied in agriculture, medicine, scientific investigations, and many industries including the treatment of drinking water. Such wide application of ion exchange process is due to two main factors: the heterogeneity of the system (e.g., a possibility of a simple phase separation like just passing a water solution through a column with the ion exchanger) and the ability of materials for ion exchange which provides a possibility to remove selectively ions from water or the separation of ions with a different charge, the value of charge or degree of hydration. The simplicity and efficiency of ion removal (either total or partial) from a solution, which could easily be realized in larger scales (from the laboratory columns to industrial filters), became a reason of its wide application in science and technology. The development of ion exchange and adsorption science has been conditioned by the requirements of different fields of practical application: the needs of agriculture, development of atomic energy production, hydrometallurgy, food and medical industries, environmental protection and water treatment. Such wide application of ion exchange processes was a result of the successful development of new ion exchange adsorbents.

We can summarize here that the development of adsorption and ion exchange sciences is based, first of all, on the success of material sciences, for example, the development of new adsorptive and/or ion exchange materials. Improved understating of the main mechanisms taking place at the interfaces of each system (the adsorbent and ion exchanger – the adsorbate) aids the total progress. Figure 5.1 shows the main types of adsorptive materials. Taking into account that the new adsorptive materials are a key moment (e.g., basis) of the progress in adsorption science, this chapter is built upon the characterization of each type of adsorptive materials (as mentioned in Figure 5.1) with some unfair attention distribution (due to experience of the author).

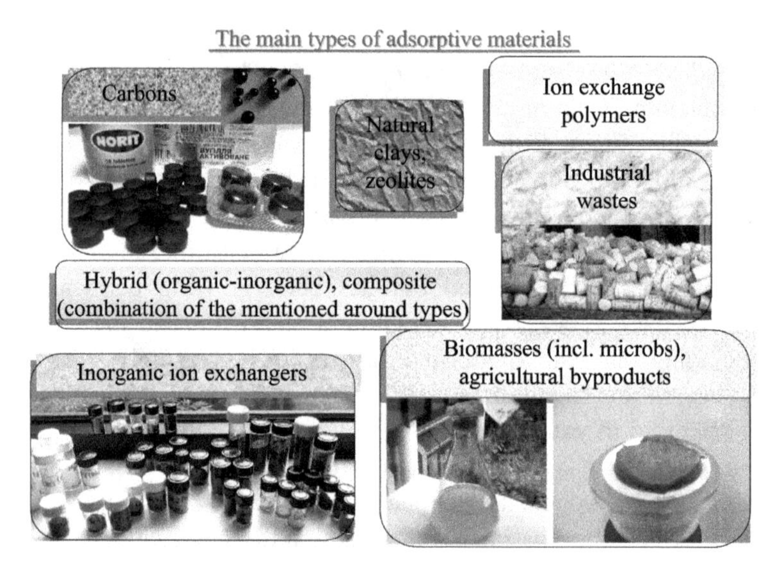

Figure 5.1 The main kinds of the adsorptive materials

5.3 CARBONS IN WATER TREATMENT AND MEDICINE

The discovery by J.T. Lovitz of the ability of carbon properties to purify water solutions from different admixtures and odors impressed the scientific community very much. Many great scientists tried to repeat the experiments with charcoal in order to find an explanation for the phenomenon of the material purifying properties. The discovery is still of great importance at the present time. Carbon adsorbents (activated, oxidized, pretreated etc.) are widely used in industries, medicine, and water treatment. Adsorption by carbons is an important part of any older (Voyutsky, 1964) or recent Colloid Chemistry book (Birdi, 2000).

Until the 20th century, carbon adsorbents (charcoals) were used mostly in the food and wine industries for the purification of water solutions. The next step in the development of carbon adsorbents was stimulated by the First World War when respirators were created due to the necessity to neutralize chemical agents. At the present time, the main area of the application of carbon adsorbents is within adsorptive purification of water solutions, separation and concentrating in gaseous and liquid phases. The role of carbon sorbents in drinking water and the treatment of waste streams has been growing and is reflected on the websites of the firms selling the materials, such as: sigma-aldrich.com/supelco. The field of carbon-based adsorbents application in medicine and pharmacueticals has been widening too.

Originally, porous sorbents were synthesized mostly by the thermal treatment of woods, and later on, of coals. Today, carbon sorbents have been produced from a variety of carbon-containing raw materials: wood and cellulose (Malikov *et al.*, 2007), coals (Drozdnik, 1997), peat (Novoselova *et al.*, 2008), petroleum and coal pitches (Pokonova, 2001), synthetic polymeric materials (Mui *et al.*, 2004), liquid and gaseous hydrocarbons (Gunter and Werner, 1997; Likholobov 2007) and different organic wastes (Long *et al.*, 2007), (*http://home.att.net/~africantech/GhIE/ActCarbon.PDF*). Three-hundred types of activated carbons have been produced by the German company DonauCarbon for different types of water (drinking, swimming pools, aquariums) and waste water purification as well as carbon for medicine and industry (food industries, atomic energy production, carriers/support for catalysts).

Granular activated carbon (GAC) is a well known cost-effective, conventional adsorbent which has been widely used since the early 1990s for water and air purification. This porous material has a structure which contains both ordered and disordered carbon rings. In contrast graphite carbons consist of free porous space which is a three-dimensional labyrinth of pores (Voyutsky, 1964; Drozdnik, 1997; Birdi, 2000; Pokonova, 2001; Mui *et al.*, 2004; Malikov *et al.*, 2007; Novoselova, 2008). Many recent studies have looked at investigation to improve the understanding of the structural characteristics of carbon adsorbents, (Barata-Rodrigues *et al.*, 1998; Gun'ko *et al.*, 2003; Gun'ko and Mikhalovky, 2004). Generally the pore sizes of carbon adsorbents are different. Traditionally, they are grouped as follows:

- micropores ($\leq$ 2 nm);
- mesopores (2–50 nm);
- macropores ($>$50 nm).

Carbons are able to sorb (take up) different molecules and ions from liquid solutions and gases due to their highly developed porous structure and their surface in functional groups. The application of carbon adsorbents and their selectivity towards contaminants that are to be removed from water solutions depends, first of all, on the chemical composition and concentration of the surface functional groups. The main functional groups of carbons are oxygen-containing groups, such as, phenol (hydroxylic), carbonyl, carboxylic, ether, and lactones which are formed during oxidizing treatment of the surface (Voyutsky, 1964; Biniak *et al.*, 1997; Rodrigues-Reinoso and Molina-Sabio, 1998; Figuiero *et al.*, 1999; Yin *et al.*, 2007; Shen *et al.*, 2008). Different conditions around the carbon surface treatment (activation agents, temperature, time, precursors, and templates) can provide the carbon surfaces with functional groups containing nitrogen, sulfur, halogens, and phosphorus.

One of the first explanations of ion exchange ability of carbon adsorbents, (which were nonpolar and, at first sight, should not sorb polar and ionic substances), was given by N.A. Shilov and reflected in the Colloid Chemistry textbook (Voyutsky, 1964). If the raw materials for carbon synthesis contain inorganic admixtures, the source of ion exchange can be explained by the presence of the cations in the carbon structure which can be easily exchanged. If inorganic admixtures are absent in the composition of the raw materials chosen for carbon activation, cation and anion exchange groups can be formed correspondingly at the relatively lower and higher temperature as showed in Figure 5.2. Intermediate temperature creates the conditions when both type of ion exchange groups (cation and anion exchange) can be formed at the surface.

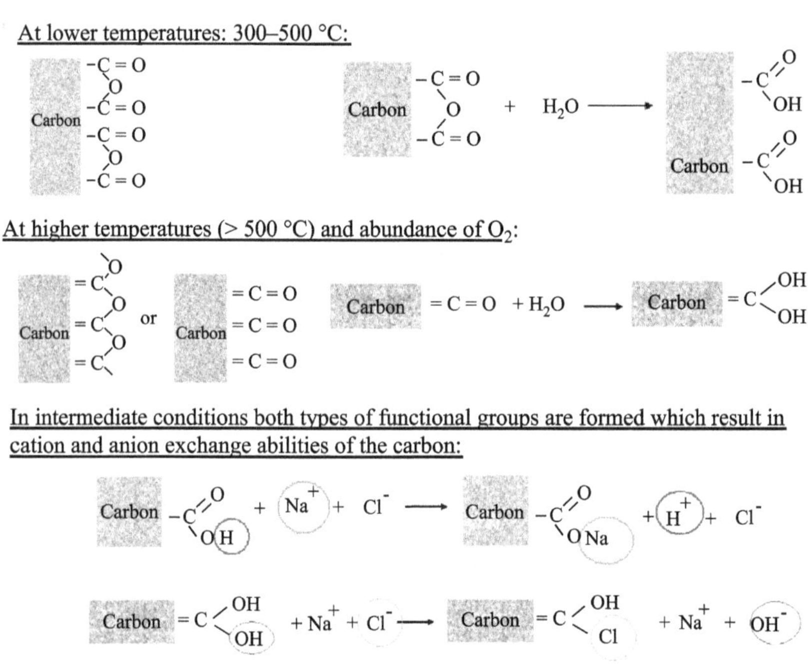

At lower temperatures: 300–500 °C:

At higher temperatures (> 500 °C) and abundance of O_2:

In intermediate conditions both types of functional groups are formed which result in cation and anion exchange abilities of the carbon:

Figure 5.2 Mechanism of formation of ion exchange groups on the surface of carbon adsorbents as suggested by N.A. Shilov (Voyutsky, 1964)

Figure 5.3 shows (schematically) the main functional groups of carbon surface responsible for binding of molecules and participles via complexation and ion exchange as shown in numerous publications (Biniak *et al.*, 1997; Rodrigues-Reinoso and Molina-Sabio, 1998; Figuiero *et al.*, 1999; Yin *et al.*, 2007; Shen *et al.*, 2008).

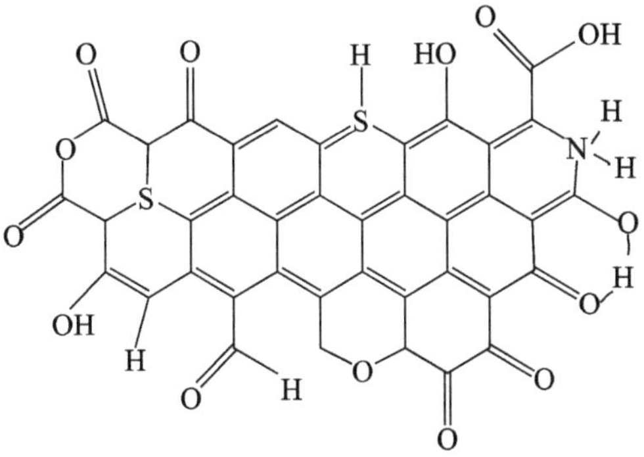

Figure 5.3 Schematic presentation of carbon surface main functional groups

A variety of the functional groups, confirmed by spectroscopy and regular chemical analyses, makes the materials selective not only towards organic contaminants (Bhatnagar and Jain, 2006; Fletcher *et al.*, 2008; Anbia *et al.*, 2009; Mansoor *et al.*, 2009) but also towards cationic (Kononova *et al.*, 2001; Park *et al.*, 2007; Lach *et al.*, 2007; Namasivayam *et al.*, 2007; Valinurova *et al.*, 2008) and anionic (Mandich *et al.*, 1998; Lach *et al.*, 2007; Namasivayam *et al.*, 2007) substances. GAC is widely used to remove different natural chemical pollutants from the air or water streams, both in the field and in industrial processes such as spill cleanup, groundwater remediation, drinking water treatment, air purification, and the capture of volatile organic compounds (VOCs).

5.4 ZEOLITES (CLAYS)

The term "zeolite" was originally coined in the 18th century by the Swedish mineralogist A.F. Cronstedt, who observed that upon heating, the stone began to "dance about" (Colella and Gualtieri, 2007). This was due to the water adsorbed inside the zeolite's pores being driven off. The name came from Greek: "zein" means "to boil", "lithos" means "a stone". We still, of course, use zeolite today to remove unwanted ammonia from pond water (discovery of English soil scientist H.M. Thompson), however, the scientific community and the water industry have discovered (and developed) a variety of zeolite-based materials including some for water purification due to the unique structure of natural zeolites (which, in addition, are very beautify crystals) and a property to exchange substances from liquids and gases.

Zeolites are crystalline aluminosilicates with a microporous framework built from corner sharing SiO_4^{4-} and AlO_4^{5-} tetrahedral (with general formula: $M_n[(AlO_2)\times(SiO_2)y]*mH_2O$) which gives a large open framework structure.

The crystals are highly porous and are veined with submicroscopic channels. A large variety of different frameworks can be produced in this way. On March 10, 2009, the International Zeolite Society assigned 191 framework type codes of zeolite materials (*http://www.iza-structure.org/*). In 2007 the latest atlas of zeolite framework types (Baerlocher *et al.*, 2007) and a collection of XRD spectra (Treacy and Higgins, 2007) were published. Due to uniformity of the pore dimensions, these materials, often called Molecular Sieves, display many advantages. They are low-cost materials, thermally ($>800°C$) and chemically stable, of uniform size, and can develop specific surface area up to 690 m^2/g (Erdem-Şenatalar and Tatlıer, 2000), non-toxic and environmentally friendly. Due to the variety of these materials, adsorptive selectivity towards different molecules and particles is also wide. Zeolites have the ability to sort molecules selectively based primarily on a size exclusion process due to a very regular pore structure of molecular dimensions, for example, only molecules of a certain size can be absorbed by a given zeolite material, or pass through its pores, while bigger molecules cannot. The net negative charge in zeolites, within the symmetrical voids hold the cations for the cation exchange capacity. Ion exchangeable ions, such as potassium, calcium, magnesium and sodium, the major cations, are held electronically within the open structure (pore space) – up to 38% void space.

A.A.G. Tomlinson (1998) explained what zeolites are (history, classification, structure, nomenclature, application): "From being mere geological curiosities one hundred years ago, zeolites have progressed to their present status as indispensable absorbents and catalysts ..." (Henmi *et al.*, 1999).

From natural zeolites, water (and other) industries went to synthesis of the artificial zeolites. The first artificial zeolite was synthesized in the 1950s. Patents describing the new methods of zeolites synthesis and application have been issued on regular basis (Henmi *et al.*, 1999; Fiore *et al.*, 2009). Besides development of new zeolite-type materials, the theory of ion exchange has also been under focus (Petrus and Warchol, 2005; Melian-Cabrera *et al.*, 2005; Gorka *et al.*, 2008). The variety of natural zeolites and their origin is a source of curiosity for many researchers (Bartenev *et al.*, 2008).

5.5 ION EXCHANGE RESINS OR ION EXCHANGE POLYMERS

The term ion-exchange resins remains very commonly used to refer to ion exchange macromolecular materials based on different organic polymers despite the strong discouragement of IUPAC (IUPAC, 2004). The materials have a highly developed structure of pores on the surface and are abundant in functional groups capable to exchange cations and anions. The main advantages of ion-exchange resins are their high mechanical and chemical stabilities in combination with high adsorptive (e.g., ion exchange) capacity. If the ion-exchange material is an organic polymer (e.g., resin) the exchange of ions can take place in the whole volume of the resin due to the free penetration of the ions into the porous structure of the resin. Therefore, the most common usage of these materials has been water softening and purification. It is also employed in purification of liquid foods (Miers, 1995). Water softening was the first industrial application to use ion exchange. The process was firstly used by Robert Gans (working for German General Electric Company) in 1905 who suggested improvements in the type of materials and equipment, as well as water softening possibilities using permutite (Helfferich, 1962). Gans's process is still one of the simplest methods for softening water. The process of water softening means passing water that contains 'hardness' ions (mostly, calcium (Ca^{2+}) and magnesium (Mg^{2+})) through a column containing a strongly acidic cation exchange resin in sodium (Na^{+}) form. The calcium and magnesium ions are exchanged for the equivalent amount of sodium ions.

Ion-exchange resins have been obtained by condensation, polymerization, or from monomers, which already contain the active functional groups, or just introduce those groups in the previously synthesized resins. The ion-exchange capacity of a resin increases with an increasing number of functional groups. On the other hand, this increase in functional groups also leads to negative effects in resin properties: the resins are then liable to swell in water and become more soluble in water. The formation of cross-linking bonds can overcome the problem of swelling and solubility in water. Figure 5.4 shows(schematically) the structure of a cation-exchange resin with the typical functional groups of cation exchangers: $-SO_3H$, $-COOH$, $-OH$. Some exchangers contain only one type of functional group and others contain a combination of a few functional groups. Figure 5.4 shows cation-exchange resins in H^{+} form where the exchangeable ions are H^{+}. For water softening, Na^{+} form cation exchanger should be used.

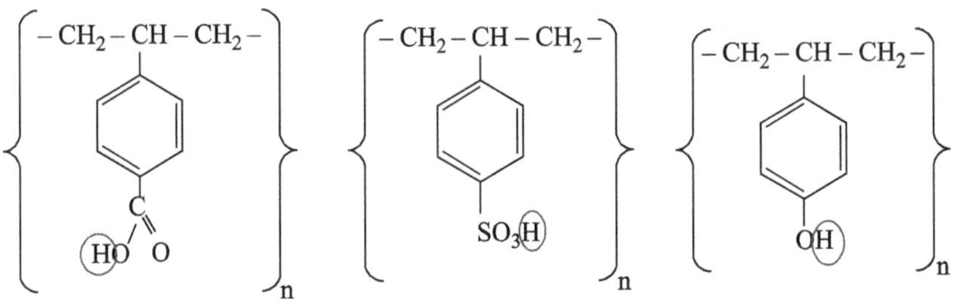

$$\left\{ \begin{array}{c} -CH_2-CH-CH_2- \\ | \\ C_6H_4 \\ | \\ C(=O)-OH \end{array} \right\}_n \quad \left\{ \begin{array}{c} -CH_2-CH-CH_2- \\ | \\ C_6H_4 \\ | \\ SO_3H \end{array} \right\}_n \quad \left\{ \begin{array}{c} -CH_2-CH-CH_2- \\ | \\ C_6H_4 \\ | \\ OH \end{array} \right\}_n$$

Figure 5.4 Repeating units of polymeric cation exchangers with the main functional groups (–SO3H, –COOH, –OH) and polystyrene matrix

The main active functional groups of anion exchangers are nitrogen-containing groups ($-NH_2$, $=NH$ and $\equiv N$) from aliphatic or aromatic amines. Figure 5.5 shows the typical structure of anion exchangers.

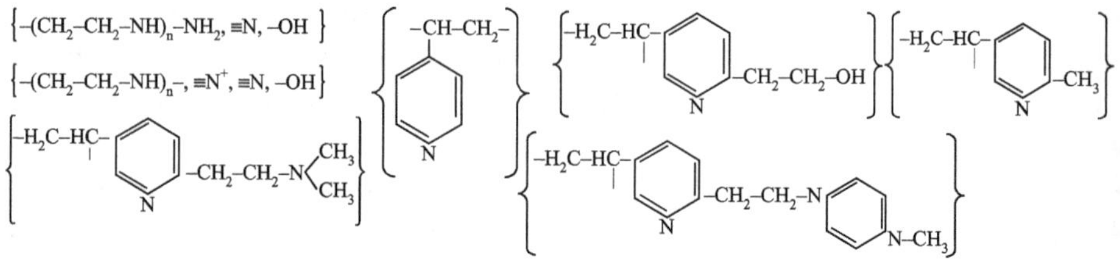

Figure 5.5 Typical repeating units of polymeric anion exchangers

Chemically active polymers have different properties: they are capable of ion exchange, complexation, reduction-oxidation and precipitation. Many colleagues divide ion-exchange resins into three groups (Sengupta and Sengupta, 1997). The first group includes nonfunctional strong acidic or strong basic groups having ion-exchange properties only. The second group collects cross-linked polymers of three-dimensional structure which show complexing or combined complexing and ion-exchange abilities due to the presence of electron-donor functional groups capable of forming complex substances with transition metals having vacant orbital. The third group includes cross-linked polymers which have the ability for redox transformation or combined ability for ion exchange and redox reactions.

The first ion-exchange resins were introduced in 40s of the last century as a more flexible alternative to natural or artificial zeolites (Irving *et al.*, 1997; Whitehead, 2007). Irving M. Abrams and John R. Milk (1997) offer a very nice brief review of the history and development of macroporous ion-exchange resins, starting with the condensate polymers in the early 1940s, continuing with the addition of polymers in the late 1950s, and evolving into new adsorbents for a variety of interesting applications. They also pay some attention to the effect of different methods of resin preparation, and on the physical properties and internal structure of the (macroporous) resins synthesized. Ion-exchange capacity depends on the type of the active functional groups of the exchanger, the chemical nature of the exchanging ions, their concentration in the solutions, and the pH of the solutions.

Since the last almost 70 years, when ion-exchange resins were brought to water treatment (and purification technologies in general), the interest in this type of materials has not weakened but has been growing, this is reflected in a variety of scientific and technological applications, the creation of a journal *Reactive and Functional Polymers*, an increasing number of commercially available ion-exchange resins (*www.water.siemens.com*; Vollmer *et al.*, 2005) and the development of new ion exchangers with nontraditional morphology (Kunin, 1982) and from various raw materials (Wayne *et al.*, 2004).

Ion-exchange resins based on styrene and divinylbenzene are now dominating the international market due to their high quality, relatively low price and easy accessibility of the raw materials for synthesis. The development of porous copolymers by the introduction of solvents during polymerization has led to a considerable increase in the number of ion exchange resins with improved kinetics of ion exchange, high osmotic stability and sieve effect (Seniavin, 1981; Kunin, 1982) and the necessity to develop regulations for using those materials in the food, water, and beverages industries (Franzreb *et al.*, 1995).

German chemists working within the group of Wolfgang H. Hoell (Institute of Technical Chemistry from Karlsruhe) have been contributing considerably to the development of ion-exchange materials, and especially the theory of ion exchange in water treatment (Franzreb *et al.*, 1995; Hoell *et al.*, 2002, 2004). Ion-exchanger polymers have remained one of the most important groups of adsorptive materials which are widely employed for the removal of anions (Karcher *et al.*, 2002; Marshal *et al.*, 2004; Matulionytė *et al.*, 2007) and cations (Elshazly and Konsowa, 2003; Vollmer and Gross, 2005; Kim *et al.*, 2007) as well as polar molecules (Fettig, 1999; Cornelissen *et al.*, 2008) in drinking water treatment. Interesting results were obtained by Matulionytė *et al.* (2007) who studied the possibility of anion extraction from fixing process rinse waters contaminated by Br^-, $S^2O_3^{2-}$, SO_4^{2-} , SO_3^{2-}, CH_3COO^- and silver thiosulfate complex anions using Cl^- form of commercial anion-exchange resins: Amberlite IRA-93 RF, Purolite A-845, Purolite A-500 and AB-17-8.

5.6 INORGANIC ION-EXCHANGERS

Stricter regulations in water quality and the reduction of the maximum contaminant levels (for new toxicological investigations) caused the necessity of looking for new adsorptive materials with much higher selectivity towards the target contaminants. Also, if ion-exchange polymers do not meet the requirements of high chemical, thermal, and radioactive stability and an alternative is required, inorganic ion-exchange adsorbents have come under the focus of researchers and civil engineers. Thus, the need for inorganic ion-exchange adsorbents has been arising mostly when conventional ion-exchange resins (polymers) do not meet the requirements of the World Health Organization (WHO) (and their maximum permissible levels) and/or the requirements for physico-chemical characteristics of the materials in definite conditions of purifying solutions (high thermal, chemical and radioactive stability). Some of the inorganic ion-exchangers can keep radioactive irradiation up to 10^9 rad and higher without any damage to the adsorbent structure (Seniavin, 1981).

The possibility of choice and synthesis of inorganic adsorbents with the desired properties is almost unlimited due to (1) a variety of different classes of lower solubility inorganic compounds which are able to take up ions from water solutions by different sorptive actions; (2) the development of synthesis methods; and (3) the process of material modification/pretreatment during and after synthesis. Currently, the information on the adsorptive properties of the widely known classes of inorganic adsorbents, such as oxides, hydrous oxides, sulfides, phosphates, alumosilicates, and ferrocyanides have been studied extensively in the literature, and new methods of their synthesis have been regularly developing. New adsorbents based on the above mentioned substances (which often exceed the ion-exchange resins on ion-exchange capacity and selectivity, and in addition are characterized by higher chemical, thermal and radioactive stability) have been produced. The main classes of inorganic ion exchange adsorbents are discussed below.

5.6.1 Ferrocyanides adsorbents

These are based on simple (such as $Zn_2[Fe(CN)_6]$) or mixed (such as $K_2Zn_3[Fe(CN)_6]_2$) ferrocyanides. In the 1960s, this group of chemical compounds was recognized as ion exchangers with high selectivity towards cesium ions (Kawamura *et al.*, 1969; Seniavin, 1981). Ferrocyanides have cation and anion exchange capacities and can serve as scavengers for toxic cations and anions, as well as organic molecules (Ali *et al.*, 2004). Sorptive properties of ferrocyanides have been used not only in purification technologies but also to explain phenomena in geochemistry. Particularities of the interaction of two naturally occurring aromatic α-amino acids (namely tryptophan and phenylalanine) with zinc, nickel, cobalt, and copper ferrocyanides has been shown by Wang *et al.* (2006), and the hypothesis that metal ferrocyanides might have concentrated the biomonomers on their surface in primeval seas during the course of chemical evolution has been suggested. Recent studies have concentrated mostly on the immobilization of insoluble ferrocyanides on the surface of zeolites, clays, carbons, and hydrous oxides. Cesium removal from deionized water, seawater, and limewater was investigated using copper ferrocyanide incorporated into porous media including silica gel, bentonite, vermiculite, and zeolite as adsorbents (Huang and Wu, 1999). A granular inorganic cation exchanger showing high selectivity to cesium (Sharygin *et al.*, 2007) was a mixed nickel ferrocyanide distributed over zirconium hydroxide as an inorganic carrier having a trademark Termoxid-35. The structure of ferrocyanides is shown in Figure 5.6.

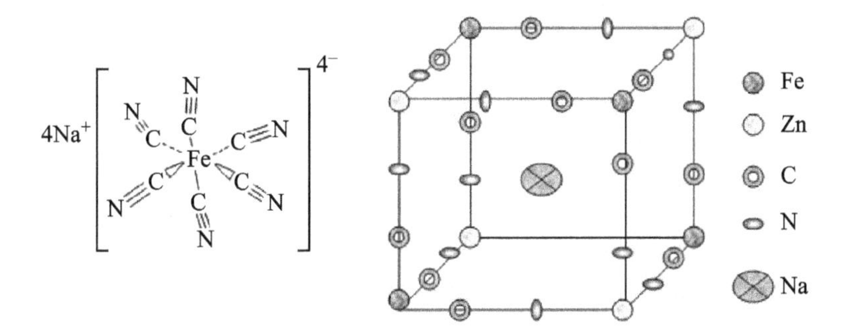

Figure 5.6 Schemes of the structure and a fragment of the crystal structure of ferrocyanide

The main sorption centers of ferrocyanides are as follows.

- Metal ions (Na^+) or H^+ existing in the structural emptiness of the crystals.
- The existence of charge due to the possible change of the oxidation state of the matrix metal or unsaturated coordination number (which can be temporarily saturated by aqua groups) due to vacancies in the matrix structure which can be filled due to sorption actions.
- Metals of the matrix can exchange cations from the solution or bind anions (including OH^-).
- The sorption of ions by ferrocyanides is linked to a few processes such as exchange for metal ions (e.g., Na^+) or H^+ existing in the structural emptiness (interstices) of the crystals, exchange with the metals ions of the matrix or non-ion exchange introduction into structural emptiness or micro-interstices with the following compensation of the excess charge or saturation of coordination number of the matrix atoms (e.g., exchange with aqua groups).

Of particular value with ferrocyanides is their exceptionally high selectivity towards ions of Cs^+, Rb^+, and Tl^+ and very little pH dependence (Kawamura *et al.*, 1969; Tananaev *et al.*, 1971; Volgin, 1979; Jain *et al.*, 1980; Clarke and Wai, 1998; Huang *et al.*, 1999; Ali *et al.*, 2004; Sharygin *et al.*, 2007). Ferrocyanides can selectively remove heavy alkaline metals and thallium from complex on the composition water solutions including strong acidic waste waters of color metallurgy and liquid radioactive wastes. Ferrocyanides have also been investigated for their uptake capacity towards aminopyridines (Kim *et al.*, 2000). Nickel ferrocyanide was found to be a better adsorbent as compared to cobalt ferrocyanide.

The other important type of inorganic ion-exchangers demonstrating both cation and anion exchange properties are based on *individual and mixed hydrous metal oxides*. The term hydrous oxides refers to all oxides (MO_x), hydroxides [$M(OH)_x$], and oxyhydroxides (MO_xOH_y) of a metal, *M*. Other terms that have been used include oxide hydrates, oxide-hydroxides, and sesquioxides among others. Hydrous oxides of Fe, Al, Mn and a combination of these metals with other metals are a predominant part of this class of inorganic ion-exchangers. The main known hydrous oxide compounds of these chemical elements are: *Goethite* ($FeO(OH)$), Hematite (Fe_2O_3), Magnetite (Fe_2O_3), *Maghemite* (Fe_2O_3, γ-Fe_2O_3), Goethite ($FeO(OH)$), *Lepidocrocite* (γ-$Fe^{3+}O(OH)$), Ferrihydrite (hydrous ferric oxyhydroxide minerals interpreted from XRD as α-Fe_2O_3 or δ-FeOOH), *Gibbsite* (γ-$Al(OH)_3$ but sometimes as α-$Al(OH)_3$), *Bayerite* (α-$Al(OH)_3$, but sometimes as β-$Al(OH)_3$), *Diaspore* (a native aluminium oxide hydroxide α-$AlO(OH)$), Hollandite ($Ba(Mn^{4+}Mn^{2+})_8O_{16}$, Todorokite (a rare complex hydrous manganese oxide mineral with formula $(Mn,Mg,Ca,Ba,K,Na)_2Mn_3O_{12} \cdot 3H_2O$), Corundum, *Pyrolusite,* Ramsdellite and Birnessite (minerals consisting essentially of manganese dioxide (MnO_2), Hausmannite (a complex oxide of manganese, $Mn^{2+}Mn_2^{3+}O_4$).

The main adsorptive active sites of these materials can be classified as follows.

- Aqua groups coordinated around metal atoms of the sorbent matrix can be exchanged with other polar molecules of the solutions (particularly, exchanged for neutral molecules and anions) due to dipole-dipole interaction.
- Surface hydroxyl groups (OH^-) can be exchanged with the solution anions which have higher coordination ability than OH^-.
- Hydroxyl groups in the bridge position: the bond forming bridges can be broken and the consequent steps are according to the possible mechanisms of adsorption.
- Oxygen atoms in the bridge or end-positions can be protonated, what lead to a break of the bridges and unsaturated coordination which will follow by binding of metal ion and anions, particularly, OH^-.
- Metal ions of the adsorbent matrix can bind aqua groups, anions including OH^- and can be exchanged directly for other metal ions of the contacting solution.
- If the structural emptiness of the matrix filled with cations or H^+, the latest can be exchanged with the ions of the contacting solution.
- Noncompensated charge or nonsaturated coordination number (due to a change of the matrix element(s) oxidation state or replacement of the matrix ions by ions of the other size can lead to binding the ions from the solution.

Cations of a solution can be bound to the surface of solid adsorbents due to: (1) exchange of hydrogen protons of protonated sorption sites; (2) exchange with metal ions (M^+) and H^+ existing in the emptiness of the crystal matrix; (3) non-ion exchange introduction into the structural emptiness and micro-emptiness with the following charge redistribution in the whole matrix using the closest environment of the emptiness; (4) direct ion exchange with the ions of the matrix; (5) binding of the ions or molecules to oxygen-containing sorption sites (contact adsorption) due to unsaturated coordination number of oxygen (such sorption sites can be formed due to preliminary uptake of oxy-anions (HSO_4^-, $H_2PO_4^-$ etc.); (6) direct biding to metal atoms of the sorbent matrix (contact adsorption).

Anions of a solution can be bound to the surface of an adsorbent due to: (1) direct exchange to the hydroxyl ions (OH^-) at the end position; (2) exchange with hydroxyl groups with the following binding to metal atoms of the matrix after a break of the bridge binds; (3) binding to metal ions of the matrix which have unsaturated coordination number; (4) exchange with the anions existing in the structural emptiness of the matrix; (5) exchange with the anionic groups of the sorbent matrix; (6) electrostatic interaction with the sorbent surface with high positive charge which can be generated due to the previous adsorption acts or due to change of the oxidation state of the matrix chemical elements.

Simultaneous uptake of cations and anions from a solution can take place due to few reasons: re-charge of the adsorbent surface due to the ion-exchange process with the following adsorption of the opposite charge ions; adsorption of the molecules such as MX_n, (where X is OH^-, Cl^- etc.) caused by the unsaturated coordination number of the atoms of the matrix or by the necessity to alternate cations and anions in the channels of the sorbent structure; uptake of cations and anions of the salts by metal hydrous oxides and formation of basic salts, and the other reasons.

The field of application of inorganic ion-exchangers based on individual and mixed hydrous oxides of metals is very broad due to the variety of possible mechanisms of sorption described above. They are explored as cation and anion exchangers.

For instance, the cation-exchange properties of manganese (hydrous) oxides were shown in the efficient adsorption of lithium, cadmium, potassium, radium etc. Manganese oxide adsorbents have been found to be very efficient in the removal of lithium from sea water (Ooi *et al.*, 1986). Later studies by the same principle investigator led to the improvement of the syntheses methods and obtaining new manganese oxide adsorbent ($H_{1.6}Mn_{1.6}O_4$) from precursor $Li_{1.6}Mn_{1.6}O_4$ by heating $LiMnO_2$ at 400°C (Chitrakar *et al.*, 2001). $LiMnO_2$ was prepared by two methods: hydrothermal and reflux. The new manganese oxides demonstrated higher adsorptive capacity (40 mg/g) as compared with other known adsorbents (Chirakar *et al.*, 2001). The cation exchange ability of manganese oxides was employed to remove cadmium (Hideki *et al.*, 2000), potassium (Tanaka and Tsuji, 1994) and cobalt (Manceau *et al.*, 1997). The addition of hydrous manganese oxide (HMO) has been identified by USEPA as an acceptable technology for radium removal in groundwater supplies (Valentine *et al.*, 1992; EPA, 2000; Radionuclides, 2004). This method of radium removal technology is particularly suited to small and medium-sized systems. This process relies on the natural affinity of radium to adsorb onto manganese oxides. Anion exchange properties of manganese oxides (natural, synthesis and biogenic) have been widely investigated with great attention to potential arsenic removal (Moore *et al.*, 1990; Manning *et al.*, 2002; Ouvrad *et al.*, 2002a, b; Katsoyiannis *et al.*, 2004), however, other anions have also been under focus (Ouvrad *et al.*, 2002a).

Arsenic contamination is widely known global problem (Wang and Wai, 2004). Exposure to high arsenic concentrations ($>100\ \mu g\ L^{-1}$) can result in chronic arsenic poisoning called Arsenicosis or Black Foot disease. The most serious damage to health has been in Bangladesh and West Bengal, India (www.sos-arsenic.net; www.who.int/water_sanitation_health/dwq/arsenic/en/). However, long-term exposure to lower arsenic concentrations, such as the old drinking water standard for arsenic ($50\ \mu g\ L^{-1}$), is also dangerous for humans as it can result in cardiovascular diseases, endocrine system disorders and can cause cancer. Therefore, the European Union (EU) and the United States established a new standard for levels of arsenic in drinking water of $10\ \mu g\ L^{-1}$. Since 1998, when this decision was made, new toxicological data suggest that this new standard is not safe either. As a result, some countries have adopted stricter arsenic guidelines for drinking water than the current WHO guidelines. In Denmark, the national guideline has already been lowered to $5\ \mu g\ L^{-1}$ (Danish Ministry of the Environment, 2007), as well as in the American state of New Jersey (New Jersey Department of Environmental Protection, 2004). In addition, the American Natural Resources Defense 15 Council (Natural Resources Defense Council, 2000) advises that the drinking water standard should be set at $3\ \mu g\ L^{-1}$. Australia has a drinking water guideline for arsenic of $7\ \mu g\ L^{-1}$ (National Health and Medical Research Center, 1996).

Activated alumina is one of the most popular adsorbents used on an industrial scale for toxic anion removal. It is obtained by thermal dehydration of aluminium hydroxide to retrieve the materials with high specific surface area and a distribution of macro- and micro-pores. Activated Al_2O_3 has been studied extensively for arsenic adsorption (Gupta and Chen, 1987; Singh and Pant, 2004; Manjare *et al.*, 2005) and has been effectively used for arsenic removal from drinking water at pH 5.5 at the Fallon Nevada, Naval Air Station (Hathaway and

Rubel, 1987). However, strong pH effect and often insufficient removal capacity of conventional activated alumina (Gupta and Chen, 1978; Manjare *et al.*, 2005; Hathaway and Rubel, 1987; Singh and Pant, 2004) conditioned the development of better modification of this ion-exchange adsorbent. To prepare an ideal adsorbent with uniformly accessible pores, a three-dimensional pore system, a high surface area, fast adsorption kinetics and good physical and/or chemical stability mesoprous alumina with a large surface area (307 m^2/g) and uniform pore size (3.5 nm) was developed and tested for arsenic removal (Kim *et al.*, 2004). The maximum As(V) uptake by mesoporous alumina was seven times higher (121 mg[As(V)]/g and 47 mg[As(III)]/g) than that of conventional activated alumina.

Hydrous oxides, oxides, oxyhydroxides of iron, amouphous hydrous ferric oxide (FeO-OH), goethite (α-FeO-OH), hematite (α-Fe_2O_3) (as well as modification by iron compounds) is another important class of substances investigated for their adsorptive affinity to arsenic (Ferguson and Gavis 1972; Wilkie and Hering 1996; Altundogan *et al.*, 2000; , Roberts *et al.*, 2004; Saha *et al.*, 2005). Amorphous Fe(O)OH had the highest adsorption capability. At first glance, people try to explain the higher uptake capacity to arsenic by the highest surface area, however, for ion exchange, as the main or initiating mechanism of adsorption, this rule (surface area – adsorptive capacity) is not directly proportional. Most iron oxides are fine powders that are difficult to separate from solution afterwards. Therefore, some technological approaches are needed to make these materials more suitable for column conditions. For instance, the EPA proposed iron oxide-coated sand filtration as an emerging technology for arsenic removal at small water facilities (USEPA, 1999; Thirunavukkarasu *et al.*, 2003). (Hydrous) oxides of iron have been also tested for their adsorptive capacity to the other anions such as phosphate, bromate (Kang *et al.*, 2003; Bhatnagar *et al.*, 2009).

The list of inorganic ion-exchangers based on metal oxides is much longer and also include zirconium (Kim, 2000) and titanium dioxides (Zhang *et al.*, 2009), cerium dioxide (Watanabe *et al.*, 2003; Bumajdad *et al.*, 2009).

In the last decade mixed hydrous oxides attracted much more interest from researchers compared to individual hydrous oxides as their fitting and changing chemical composition gives more options to change the structure of the materials and to achieve the desired adsorptive properties and selectivity toward target ions and molecules. It was noticed that adsorption of anions onto mixed hydrous oxides of Fe and Al was less dependent on pH compared to individual ones (Chubar *et al.*, 2005a, b, 2006). Venkatesan *et al.* (1996) were trying to establish the correlation between the surface properties of a silica-titania mixed hydrous oxide gel and adsorptive properties towards radioactive strontium.

Tomoyuki *et al.* (2007) synthesized ten Si-Fe-Mg mixed hydrous oxide samples with changing Si, Fe(III) and Mg molar ratio, detected the influence of the percentage of Mg on the structure of the materials and studied adsorption of arsenite, arsenate and phosphate on those materials.

Inorganic ion-exchangers based on *layered mixed hydrous oxides and hydrotalcite* type materials gained the reputation of being very competitive adsorptive materials capable of solving many tasks of purification technologies. Bruna *et al.* (2009) discovered that hydrotalcite-like compounds ([$Mg_3Al(OH)_8$] Clx4H_2O; [$Mg_3Fe(OH)_8$]Clx4H_2O; Mg3$Al_{0.5}Fe_{0.5}(OH)_8$]Clx4H_2O (layered double hydrous oxides) and the calcined product of [$Mg_3Al(OH)_8$]Cl 4H_2O, $Mg_3AlO_{4.5}$ (hydrotalcites)) had a lot of potential as adsorbents of the herbicide MCPA [(4-chloro-2-methylphenoxy) acetic acid]. Sparks *et al.* (2008) developed new sulfur adsorbents (derived from layered double hydroxides) employed for COS (carbonyl sulfide) adsorption studies. It was found that selectivity of the layered mixed hydrous oxides towards arsenate was much higher as compared with balk double hydrous oxides (Chubar *et al.*, 2006). The reason for the layered materials higher selectivity to tetrahedral anions is the structure of the adsorbents (Figure 5.7) and the structural correspondence factor leading the adsorption process.

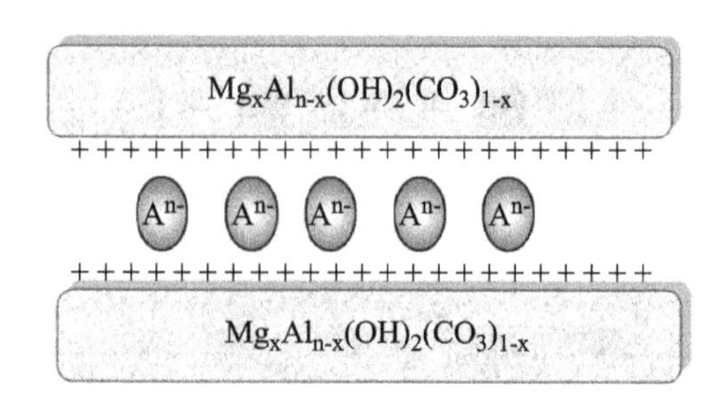

Figure 5.7 Schematic structure of hydrotalcite type substance (of general formula: Mg (Zn)2+6 Al3+2 (OH)16(CO3)) with adsorbed anion in the interlayer space

Natural (Hall and Stamatakis, 2000) and artificial (Frost *et al.*, 2005) hydrotalcites have been investigated for varying parameters for synthesis (Palmer *et al.*, 2009) and choosing the most advanced method of synthesis (in this case: sol-gel) (Lopes *et al.*, 1996) to develop novel hydrotalcite-type materials.

Inorganic ion-exchangers based on mixed and individual metal sulfides are a type of adsorbents with the chemical formula: $M_xE_yS_z$, where M and E are metals (for instance: ZnS, CuS or mixed: Zn_x Cd_{1-x} S, Cd_x Mn_{1-x} S, Mn_x Zn_{1-x} S etc.). Sulfides differ from oxides as they are larger and have the stronger ability of S, Se, and Te for polarization as compared with oxygen, which leads to increased covalent bond formation (Breg and Klarinsgbull, 1967). Atoms of Se, Se, and Te are larger than metal atoms, therefore, the crystalline structure of these materials are characterized by strong packing when formation of the emptiness is almost impossible. Thus, the mechanism of particle adsorption by sulfides excludes participation of the structure emptiness. There is a similarity between hydrous oxides and sulfides. The role of aqua and hydroxyl groups can be executed by H_2S, HS^- and atoms of sulfur. Increased concentration of H_2S and HS^- ions on the surface of metal sulfides was defined experimentally (Seniavin, 1981). These groups participate in the process of interaction with the sulfide surface. There are many similarities in the adsorption mechanisms on metal oxides and sulfides but in the latter, adsorption is a pure surface process. Some recent studies focus on new synthesis methods for metal sulfides which would advance the application of the material. Yamamoto *et al.* (1990) developed the new organosols of nickel sulfides, palladium sulfides, manganese sulfide, and mixed metal sulfides, and studied their use in the preparation of semiconducting polymer-metal sulfide composites. Manolis *et al.* (2008) reported on the family of robust layered sulfides $K_{2x}Mn_xSn_{3-x}S_6$ ($x = 0.5–0.95$) (called KMS-1) which showed outstanding preference for strontium ions in highly alkaline solutions containing extremely large excesses of sodium cations as well as in an acidic environment where most alternative adsorbents with oxygen ligands are nearly inactive. They suggest the implication of these simple layered sulfides for the efficient remediation of certain nuclear wastes. The reason of the high selectivity to strontium was a feature hexagonal $[Mn_xSn_{3-x}S_6]^{2x-}$ slabs of the CdI_2 type that contain highly mobile K^+ ions in their interlayer space that are easily exchangeable with other cations and particularly strontium. A search for a cost-effective synthesis method brought Bagreev and Bandosz (2004) to the development of efficient hydrogen sulfide adsorbents obtained by the pyrolysis of sewage sludge derived fertilizer modified with spent mineral oil.

5.6.2 Synthesis of inorganic ion exchangers

Adsorptive and ion exchange properties of inorganic ion-exchangers depend on their composition and crystalline structure particularities. Technological characteristics of the materials are defined by their shape: spherical, grains of different shape, thin layers on the surface of the inert materials, membranes, fibers, cores, tubes, porous structures etc. Desired adsorptive and technological properties of inorganic ion-exchangers can be achieved by different synthesis methods. Therefore, the predominant attention of the researchers is given to the development of new synthesis methods and the improvement of existing ones. The main way of the obtaining inorganic adsorbents includes two stages which are often consequent stages: (1) synthesis of chemical compounds (or composition) or modification of the previously developed materials; and (2) preparation of the adsorbents in a shape suitable for technological application. Synthesis of inorganic adsorbents can be run as homogeneous precipitation from the solutions and hetero-phase reactions in the systems: solid – gas, solid – liquid and just solid phase reactions. Preparation of the materials in the desired shape can be carried out by:

(1) drying the synthesized material on the surface with definite shape;
(2) blending;
(3) spray drying;
(4) freezing;
(5) pressing as tablets;
(6) making grains by few methods;
(7) using binding materials;
(8) impregnation;
(9) precipitation inside of porous materials;
(10) making spherical particles by drop-wise methods including sol-gel method.

Amphlett (1964) and Clearfield (1982) gave detailed explanations on many basic methods of inorganic ion-exchanger syntheses. The last 50 years resulted in the discovery of a variety of ion-exchange adsorbents with high selectivity to cations and anions (Bengtsson *et al.*, 1996; Bortun *et al.*, 1997; Chubar *et al.*, 2005a, b, 2006; Liu *et al.*, 2006; Zhang *et al.*, 2007; Chen *et al.*, 2009; Zhuravlev *et al.*, 2006).

When searching for the most advanced synthesis approaches, scientists often look at the sol-gel method. The advantages of materials developed via this approach are (1) physically and chemically pure and uniform particles; (2) exceptional mechanical properties owing to their nanocrystalline structure; (3) homogeneous material, since mixing takes place on the atomic scale; and (4) low processing (often ambient) temperature. The sol-gel method of synthesis was first discovered in the late 18th century and has been extensively studied since the early 1930s. Interest in this technique was renewed in the early 1970s when monolithic inorganic gels were formed at low temperatures and converted to glass without a high-temperature melting process (www.psrc.usm.edu/mauritz/solgel.html). Through this process, homogeneous inorganic oxide materials with desirable properties of hardness, optical transparency, chemical durability, tailored porosity, and thermal resistance, can be produced at room temperatures, as opposed to the much higher melting temperatures required in the production of conventional inorganic glass. Traditional precursors for sol-gel syntheses are metal alkoxides. The most widely used metal alkoxides are the alkoxysilanes, such as tetramethoxysilane (TMOS) and tetraethoxysilane (TEOS). However, other substances such as aluminates, titanates, and borates are also commonly used in the sol-gel process, often mixed with TEOS. Traditional sol-gel synthesis can not be said to be environmentally friendly, cost-effective and technologically attractive due to the expensive and toxic precursors used to run the reactions. To overcome these disadvantages researchers have been trying to develop new methods of sol-gel synthesis which avoid using toxic and expensive alkoxides as raw materials. Thus, nontraditional sol-gel synthesis methods were developed that resulted in obtaining new cation (Bortun and Strelko, 1992; Zhuravlev *et al.*, 2002, 2004, 2005) and anion (Chubar *et al.*, 2005a, b, 2008a) exchange adsorbents. The methods employed only simple inorganic metal salts, mineral acids and bases for synthesis. There is no common synthesis approach suitable for all developed materials due to the unique properties of each chemical element. Fine inorganic synthesis reactions should be developed for each new adsorbent. The main approaches of these inorganic syntheses are: (1) preliminary synthesis of precursors; (2) partial neutralization of metal inorganic slats; (3) choosing the best fitting slat for each reaction (e.g., chlorides, sulfates, nitrates of Al, Fe(II), Fe(III), Zn, Mg, Mn, Zr); (4) choosing parameters for synthesis (e.g., metal salt concentration, temperature, pH, mixing regime); (5) using some additives (organic or inorganic); (6) choosing the best alkaline agent. Examples of the final materials (A) and the precursors (hydrogels) (B) are shown in Figure 5.8. Adsorptive properties of ZrO_2 were tested by Chubar *et al.* (2005b) and Meleshevych *et al.* (2007) (Figure 5.8a). The high affinity to arsenic of the recently developed layered mixed hydrous oxide of Mg-Al has been described in the EU provisional patent application (submitted by Utrecht University) dated September 14, 2009 (Figure 5.8b).

Figure 5.8 Sol-gel generated hydrous oxides: (a) ZrO2 (Chubar *et al.*, 2005b; Meleshevych *et al.*, 2007) and (b) intermediate product (hydrogel) of the mixed layered hydrous oxide of Mg and Al (Chubar not publ.)

The materials shown at Figure 5.8 were tested for arsenate adsorption. Isotherms of $H_2AsO_4^-$ by these ion exchangers (ZrO_2 and one of the ion exchangers prepared from hydrogel as shown in Figure 5.8b) are shown in Figure 5.9 (Chubar, not publ.). The isotherms were obtained at pH = 7 (the typical pH of treated drinking water) using background electrolyte 0.1 N NaCl. The concentration of solids (adsorbents) was 2 g_{dw}/L. Batch sorption experiments were running for 72 hours. New sol-gel generated hydrous oxide of Mg-Al demonstrated very competitive removal capacity (plateau of the isotherm) which reached 180 mg[As]/g_{dw} and exceptionally high affinity to arsenate (the plot is very close to axis Y). The latter means that the material should show very high selectivity at the lower (ppb) concentrations of this target ion and should be able to meet the very stringent forthcoming requirements of WHO (5 ppb).

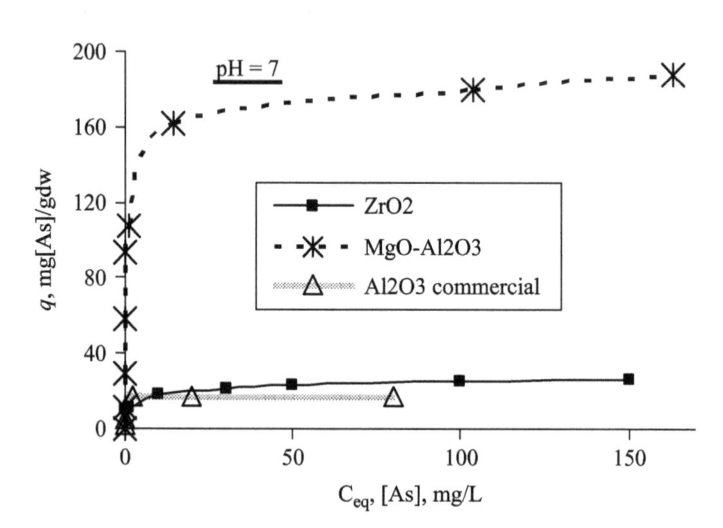

Figure 5.9 Isotherms of H2AsO4$^-$ adsorption by ZrO2 and MgO-Al2O3 hydrous oxides (shown in Fig. 5.8). Conditions of the experiment: background electrolyte: 0.1 N NaCl, adsorbent concentration: 2 gdw/L, contact time: 72 hours, temperature: 22±2°C (Chubar not publ.)

5.7 BIOSORBENTS (BIOMASSES): AGRICULTURAL AND INDUSTRIAL BY-PRODUCTS, MICROORGANISMS

For the last more then two decades biosorption or concentrating of the substances by sorbents of the natural origin has attracted the attention of specialists working in the field of treatment technology. Biomasses, which in the most cases are waste (or side products) of industry or agriculture, have a variety of functional groups and often show high sorptive selectivity toward different substances to be removed from water solutions. The main advantage of these adsorptive materials is their lower costs. Additional attractive features of biomasses (such as: competitive removal performance, frequent selectivity of the materials to target xenobiotics, a possibility to improve adsorptive properties by different ways of pre-treatment – modification of biomasses surfaces, in principle the possibility to regenerate the biomaterials (if necessary)) makes this type of adsorptive material attractive for technological application. Thus, the main pilot tasks of biosorption are: (1) to search for new capable materials suitable for industrial application; and/or (2) the pre-treatment and modification of the already found biomaterials in order to make the adsorbents capable to solve the treatment technology problem.

The origin of biosorption science is dated to 1986 when a meeting was organized by the Solvent Engineering Extraction and Ion Exchange Group of the Society of Chemical Industry in the UK and biosorption was regarded as an emergent technology (Apel and Torma, 1993). Since that time a number of research groups have been appointed to define the best biosorbents. The attractive idea has been the utilization of industrial wastes and agricultural by-products in waste treatment. Numerous biomasses have been tested for removal capacity to, first of all, heavy metals. They include industrial large-scale fermentations (antibiotic enzymes, organic acid production processes etc.) (Volesky 1990a, b, c, 1995; Wang *et al.*, 2008), agricultural by-products (peat, cotton waste, rice husk, olive pomace, pectin-reach fruit wastes etc.) (Apel and Torma 1993; Ajmal *et al.*, 2000), algae and microorganisms (seaweeds, bacteria etc.) (Kogtev *et al.*, 1996; Patzak *et al.*, 2004; Cochrane *et al.*, 2006; Chubar *et al.*, 2008b), cork tree biomass (waste of bottle stoppers from the wine industry) (Annadurai *et al.*, 2003; Chubar *et al.*, 2003a, b, c, 2004), pectin-rich fruit waste (Schiewer *et al.*, 2008) and many others. Patents were also approved for many biosorbents developed using cheap biomass as raw materials (Kogtev *et al.*, 1996). Loredana Brinza (www.soc.soton.ac.uk/BIOTRACS/biotracs) characterized the list of biomasses for their uptake ability to heavy metal ions looking for the highest affinity to the metals: Macro-algae (Brown algae: *Ascophyllum nodosum*; *Fucus vesiculosus*; *Sargassum* spp. (numerous), *Laminaria digitata*; *Laminaria japonica*; *Ecklonia sp.*; *Padina pavonia*; *Petalonia fascia*; *Pilayella littoralis*, Red algae: *Corallina officinalis*; *Gracilaria fischeri*; *Porphyra columbina*, Green algae: *Cladophora crispata*; *Codium fragile*; *Ulva fascia*; *U. lactuca* and Micro-algae (Red algae: *Porphyridium purpureum*, Green algae: *Chlamydomonas reinhardtii*; *Chlorella salina*; *C. sorokiniana*; *C. vulgaris*; *Scenedesmus abundans*; *S. quadricauda*;

S. subspicatus; *Spirogyra spp.*, Diatoms: *Cyclotella cryptica*; *Phaeodactylum tricornutum*; *Chaetoceros minutissimus*, Cyanobacteria: *Lyngbya taylorii*; *Spirulina spp*.

Biosorbent contain a variety of functional groups capable of binding metal cations: carboxyl, imidazole, sulphydryl, amino, phosphate, sulfate, thioether, phenol, carbonyl, amide and hydroxyl moieties (Volesky 1990a, b, c, 1994, 2001, 2003, 2007; Chubar *et al.*, 2008b). They were very well characterized by Wang (2009). Most of them are similar to the main functional groups of carbons and ion exchange polymers shown in Figures 5.3–5.5.

Taking into account the similarity of the main surface functional groups of biomaterials, the methods which are traditionally used for the characterization of surface chemical properties of carbons, natural zeoliets and ion-exchange resins, can be successfully employed to study surface chemical properties of biomasses. Potentiometric titration (which shows relative cation and anion exchange capacities if the data are treated correctly and if the research has been conducted in batch conditions), Boehm method (which allows us to define the concentration of strong acidic carboxylic groups, weak acidic plus lactone groups and phenolic groups), spectroscopy (FTIR) and electrophoretic mobility measurements are very useful to characterize the surface chemical properties of cork tree biomass (Chubar *et al.*, 2003c, 2004; Psareva *et al.*, 2005) and even to detect the difference in the surface chemical properties of living (viable) and dead (autoclave inactivated) gram-negative microorganism *Shewanella putrefaciens* (Chubar *et al.*, 2008b). Living *Shewanella putrefaciens* showed a capacity to sorb Mn(II) ions over a 1 month period. Formation of Mn-containing precipitates and polymeric sugars accompanied the uptake process (Figure 5.10). Complementary spectroscopy techniques (FTIR, EXAFS, and XANES) and SEM allowed the characterization of Mn-containing mineral precipitates synthesized by viable cells as a function of temperature (range 5–30°C), the contact time (up to 20–30 days), the metal loading and the bacteria density (2–4 g_{dw}/L). These parameters predetermined whether only one or a few Mn-containing precipitates in different ratio were produced. $MnPO_4$, $MnCO_3$ and MnOOH were the main precipitates formed by this bacterium (Chubar *et al.*, 2009).

Pretreatment of biomasses (surface modification) of the original biomass in order to improve or change their sorptive capacity and/or sorptive affinity towards the particles of interest is another important part of the current research of the biosorption scientists. Chemical, physical, biological, or mechanical treatment of the original biomass or a combination of the mentioned methods is usually used by the researchers. The most widely spread methods of biomass pretreatment are: (1) chemical (using acids under different concentration, different temperatures, pressure; using bases (alkali, ammonia, lime, carbonates etc), oxydative agents (oxygen, manganate, chromate, chlorates etc.), using different complexing agents; (2) physical (thermal, hydrothermal treatment, pressure, temperature etc or even production of carbons); (3) biological (enzyme and microbiological treatment). Different methods of biomass treatment are described by Walt (2003) and Kumar (2009). Patents have been also pending on the new methods of biomass treatment (Holtzapple and Davison, 1992; Hennessey *et al.*, 2009).

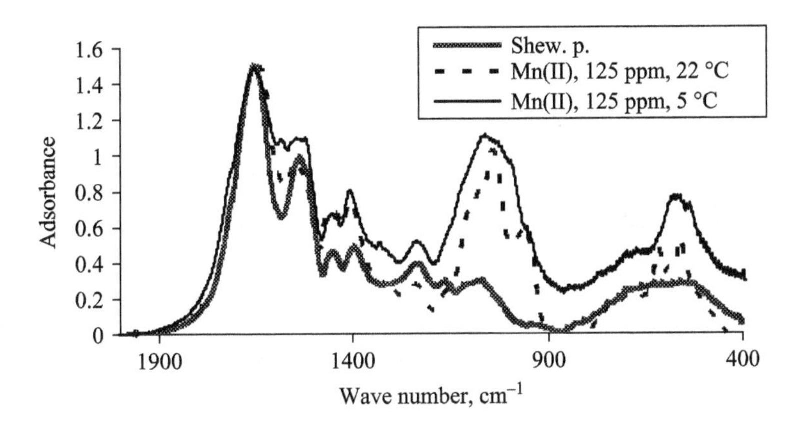

Figure 5.10 FTIR spectra of the (initially) viable Shewanella putrefaciens before and after sorption of Mn^{2+}. Batch conditions of Mn(II) sorption: contact time: 24 days, background electrolyte: 0.1N NaCl, temperatures: 5 and 22 ± 2°C, bacteria concentration: 2 g_{dw}/L (Chubar *et al.*, not publ.)

Figure 5.11 shows an increase of Cu^{2+} sorption: (a) after treatment of cork biomass with NaClO at the initial concentration of Cu(II) of 200 and 300 mg/L and (b) of carbons produced from cork biomass at different conditions at the initial concentration of Cu(II) of 10 and 200 mg/L. Careful consideration is usually needed (taking into account biomass losses and the pretreatment costs) to conclude if a pretreatment is reasonable and cost-effective.

If an original and pretreated biomasses have been already characterized for their surface chemical and adsorptive properties, the next research steps should be the same as for any other type of adsorbents: sorption studies in column conditions, investigation of a possibility for regeneration, granulation of the biomasses showing the best performance and, finally, building a technological scheme for possible biosorption implementation using packed-bed columns for biosorption and desorption as was done by Davis *et al.* (2003) and Walsh, 2008.

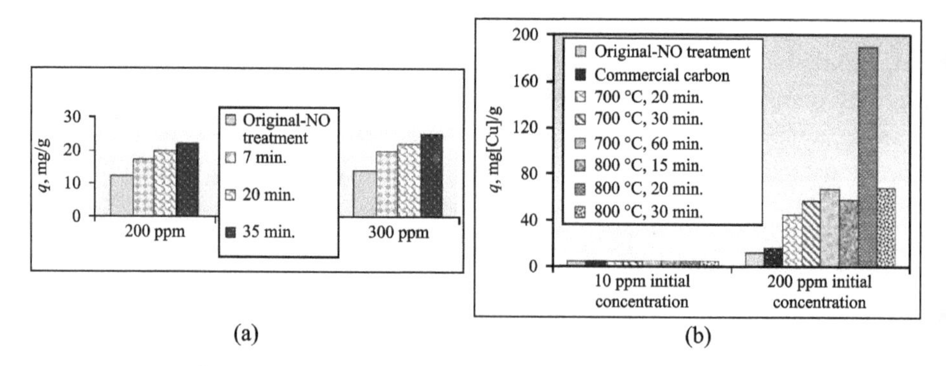

Figure 5.11 Cu^{2+} sorption on the original (not pretreated) cork tree biomass and (a) pretreated with NaClO biomass and (b) on carbons made from cork tree biomass at different temperatures (Chubar *et al.*, 2004)

5.8 HYBRID AND COMPOSITE ADSORBENTS AND ION EXCHANGERS

Hybrid and composite adsorptive materials are the most recent and most fashionable field of researches in adsorption science and technology. There are two justifications for this situation: (1) the necessity to satisfy increasingly strong WHO maximum permissible levels and look for the materials which would combine surface chemical properties of a few types of adsorbents; and (2) scientific curiosity.

In spite of the presence of the theoretical works (mostly based on mathematical modeling), which are trying to explain how to design hybrid materials on the scientific basis, an exact definition of hybrid materials (Ashby and Bréchet, 2003) is still absent. The term (hybrid) came from biology (particularly genetics). *Encyclopedia Britannica* writes: "The term hybrid ... usually refers to animals or plants resulting from a cross between two races, breeds, strains, or varieties of the same species. There are many species hybrids in nature (in ducks, oaks, blackberries, etc.), and, although naturally occurring hybrids between two genera have been noted, most of these latter result from human intervention." (*www.britannica.com/*). It is also pointed that "... the process of hybridization is important biologically because it increases the genetic variety ... which is necessary for evolution to occur". *Cambridge Dictionary Online* gives a more brief definition: "a plant or animal that has been produced from two different types of plant or animal, especially to get better characteristics, or anything that is a mixture of two very different things" (*http://dictionary.cambridge.org/*). Other sources suggest similar definitions. We can conclude here that the key elements which can be used to give a definition of hybrid material, are: (1) new material with new properties; (2) made from two completely different origin materials; (3) appears due to "human intervention", that is, made by human beings; and (4) important for the society needs, that is, "... necessary for evolution to occur".

Hybrid adsorbents have been synthesized from two types of adsorbents in different possible combinations: carbons, ion-exchange polymers, zeolites and clays, inorganic ion exchangers and natural biomasses. The most advanced methods of synthesis typical for each main type of adsorptive material are usually chosen by the researchers. Recent developments in the field of hybrid and combined adsorbents were collected in a book edited by Loureiro *et al.* (2006). Numerous publications and patent storms regularly occur in journals and online. Blaney *et al.* (2007) reported the new anion exchanger for phosphate removal. Mrowiec-Białoń *et al.* (1997) applied the sol-gel method to develop an effective hybrid adsorbent for water vapor. A combination of the sol-gel processing and molecular imprinting approach were applied by Quirarte-Escalante *et al.* (2009) to develop the hybrid adsorbent with high removal capacity to lead. Many patents and patent applications are filed in the area of hybrid adsorbents (Misra and Genito 1993; Chang *et al.*, 2007).

5.9 COMMENTS ON THE THEORY AND FUTURE OF ADSORPTION AND ION-EXCHANGE SCIENCE

The term adsorption was introduced by Kayser in 1881 to describe the increase in concentration of gas molecules in neighboring solid surfaces, a phenomenon noted by Fontana and Scheele in 1777 (Gregg *et al.*, 1982). Kayser suggested the first empirical equation of adsorption isotherm ($V = a + bP$) and introduced the term adsorption. The term adsorption isotherm was introduced by Ostwald in 1885. McBain was the first one who separated the phenomena of adsorption and absorption in 1909 (Kiefer *et al.*, 2008). The next stages of the development of the theoretical basis of the adsorptive phenomena are widely known. At the end of the last century Gibbs developed the basis of thermodynamic theory of surface phenomena which is still officially called up as classical thermodynamic theory of heterogeneous systems. The first (famous) theoretical equation of adsorption isotherm was derived by Langmuir in 1916 which has been used actively by researchers up until today. Brunaer, Emmett and Teller used a derivation of the Langmuir isotherm to derive their BET adsorption isotherm (1938) also known as Brunaer-Emmett-Teller (BET) adsorption isotherm. Many famous names of scientists, who contributed considerably to the theoretical development of adsorption science, can be still mentioned. Their names and the main equations are summarized by Dabrowski (2001). However, the modern theory of adsorption is still developing.

It is impossible to predict a far future foradsorption science. The most serious "danger" on the way is new unexpected discoveries and (micro)revolutions which can cause multidimensional consequences. One examples of such a micro-revolution is the development of carbon-carbon composite materials composed from graphitized fibrous carbon (Buckley and Edie, 1993). The second micro-revolution is likened to the discovery of fullerene which is a special modification of carbon: a carbon allotrop composed entirely of carbon in the form of a hollow sphere (buckyball), an ellipsoid, a tube (nanotubes), or a plane (graphene) (Margadonna, 2008). It is difficult to make prognoses for the future of these materials even today.

The near future prognosis of adsorption science is pretty optimistic, first of all, due to the success of adsorption material science and the broadening variety of recently developed adsorptive materials and the considerably growing application of adsorptive materials in traditional fields and in new areas. New materials include relatively expensive adsorbents for catalysis, medicine, multifunctional materials as well as cheaper materials for domestic and industrial waste streams purification including biosorbents produced from industrial or agricultural waste for one time use. The promising recently developed materials are new classes of porous materials based on carbon, mineral, metallic, polymer matrix and their composite; new types of porous materials with strictly given pore distribution on radius, wide application of block constructions including on the composite carrier etc.

Fundamental scientific tasks of the near future for adsorption science are (1) the development of a modern theory of adsorption; and (2) a theory of controlled synthesis of porous materials, formulation of their structure and texture. Wide involvement of spectroscopy (including synchrotron based X-ray techniques), quantum-chemistry modeling, mathematic modeling of the adsorption processes and electron microscopy (Roddick-Lanzilotta *et al.*, 2002; Lefevre, 2004; Chubar *et al.*, 2009) is a reliable direction to improve understanding of the main processes taking place at the interface (adsorbent – adsorbat) and to develop a modern theory of adsorption and controlled syntheses of the adsorptive materials.

Wang and Chen (2009) think that the future of biosorption science and technology is facing great challenges. Many investigators think that the failure of the commercialization process is due to mainly nontechnical pitfalls involved in the commercialization of technological innovation. For the future of biosorption, it seems there are three trends to watch. One of them is using biosorbents in a hybrid technology as one of the adsorptive materials. The second (which would lead to using living microbial cells in the industrial scale treatment plants) is to improve understanding of the processes taking place at the interface: living microorganisms – metal ions (Chubar *et al.*, 2009) and to find the conditions where anionic target species (not only metal cations) can be also removed by living or nonliving microbs (Chubar *et al.*, 2008b). It was shown that living cells of *Shewanella putrefaciens* are capable to remove Mn^{2+} over month period and to form bioprecipiatates (or new biosorbents). The last trend is to develop a good commercial biosorbent just like a kind of ion-exchange resin, and to exploit the market with great endeavor (Volesky, 2007).

5.10 ACKNOWLEDGEMENT

King Abdullah University of Science and Technology (www.kaust.edu.sa) Center-in-Development Award to Utrecht University: www.sowacor.nl (award No. KUK-C1-017-12) is gratefully acknowledged.

5.11 REFERENCES

Abrams, I. M. and Milk, J. R. (1997) A history of the origin and development of macroporous ion-exchange resins. *Reactive & Functional Polymers* **35**, 7–22.

Ajmal, M., Rao, R. A. K., Ahmad, R. and Ahmad, J. (2000) Adsorption studies of *Citrus reticulata* (fruit peel of orange): Removal of Ni(II) from electroplating wastewater. *J. Hazard. Mater.* **79**(1–2), 117–131.

Ali, S. R. and Alam, T. Kamaluddin (2004) Interaction of tryptophan and phenylalanine with metal ferrocyanides and its relevance in chemical evolution. *Astrobiology* **4**, 420–426.

Altundogan, H. S., Altundogan, S., Tumen F. and Bildik, M. (2000) Arsenic removal from aqueous solutions by adsorption on red mud. *Waste Man.* **20**(8), 761–767.

Amphlett, C. B. (1964) *Inorganic Ion Exchangers*. Elsevier Pub. Co. Amsterdam, New York.

Anbia, M. and Ghaffari, A. (2009) Adsorption of phenolic compounds from aqueous solutions using carbon nanoporous adsorbent coated with polymer. *Applied Surface Science* **255**(23), 9487–9492

Annadurai, G., Juang, R. S. and Lee, D. J. (2003) Adsorption of heavy metals from water using banana and orange peel. *Water Sci. Technol.* **47**, 185–190.

Apel, M. L. and Torma, A. E. (1993) Immobilization of biomass for industrial application of biosorption. In *Biohydromettalurgical Technologies* (eds Torma, A. E., Apel, A. E. and Vrierley, C. L.) The Minerals, Metals and Materials Society, TMS Publication, Wyoming, USA, vol. II, pp. 25–33.

Ashby, M. F. and Bréchet, Y. J. M. (2003) Designing hybrid materials. *Acta Materialia*, **51**(19), 5801–5821.

Baerlocher, C., McCusker, L. B. and Olson, D. H. (2007) *Atlas of Zeolite Framework Types*, (revised edition). Elsevier, Amsterdam.

Bagreev, A. and Bandosz, T. (2004) Efficient hydrogen sulfide adsorbents obtained by pyrolysis of sewage sludge derived fertilizer modified with spent mineral oil. *Enviro. Sci. & Tech.* **38**(1), 345–351.

Barata-Rodrigues, P. M., Mays, T. J. and Moggridge, G. D. (2003) Structured carbon adsorbents from clays, zeolites and mesoporous aluminosilicate templates. *Carbon* **41**, 2231–2246.

Bartenev, B. K., Belchinskaya, L. I., Zhabin, A. B. and Khodosova, N. A. (2008) The approach to study the sorptive ability of mineral compounds as a function of their composition. *Vestnik of Voronesh University* **2**, 133–137 (in Russian).

Bengtsson, G. B., Bortun, A. I. and Strelko, V. V. (1996) Strontium binding properties of inorganic ion exchangers. *Journal of Radioanalytical and Nuclear Chemistry* **204**(1), 75–82.

Bhatnagar, A. and Jain, A. K. (2006) Column studies of phenols and Dyes removal from aqueous solutions utilizing fertilizer industry waste. *International J. Agr Res.* **1**(2), 161–168.

Bhatnagar, A., Choi, Y. H., Yoon, Y., Shin, Y., Jeon, B-H. and Kang, J-W. (2009) Bromate removal from water by granular ferric hydroxide (GFH). *J. Hazard. Mater.* **170**(1), 134–140.

Biniak, S., Szymanski, G., Siedlevski, J. and Swiatkovski, A. (1997) The characterization of activated carbons with oxygen and nitrogen surface groups. *Carbon* **35**, 1799–1810.

Birdi, K. S. (2000) *Surface and Colloid Chemistry: Principles and Applications*. CRC Press, Boca Raton, FL.

Blaney, L. M., Cinar, S. and SenGupta, A. K. (2007) Hybrid anion exchnager for trace phosphate removal from water and waste waters. *Water Res.* **41**, 1603–1613.

Bortun, A. I. and Strelko, V. V. (1992) Synthesis sorption properties and application of spherically granulated titanium and zirconium hydroxophosphates. In *Proc. of the 4th Intern. Conf. on Fundamentals of Adsorption*, Kyoto, 58–65.

Bortun, A. I., Bortun, L., Clearfield, A., Jaimez, E., Villa-García, M. A., García, J. R. and Rodríguez, R. (1997) Synthesis and characterization of the inorganic ion exchanger based on titanium 2-carboxyethylphosphonate, *Jour. Mater. Res.* **12**, 1122–1130.

Breg, U. and Klarinsgbull, G. (1967) *Crytalline Structure of Mineral*. Moscow: Mir.

Bruna, F., Celis, R., Pavlovic, I., Barriga, C., Cornejo, J. and Ulibarri, M. A. (2009) Layered double hydroxides as adsorbents and carriers of the herbicide (4-chloro-2-methylphenoxy)acetic acid (MCPA): systems Mg-Al, Mg-Fe and Mg-Al-Fe. *J Hazard Mater.* **168**(2–3), 1476–81.

Buckley, J. D. and Edie, D. D. (1993) *Carbon-Carbon Materials and Composites*. William Andrew Publishing/Noyes.

Bumajdad, A., Eastie, J. and Mathew, A. (2009) Cerium oxide nanoparticles prepared in self-assemled systems. *Adv.Colloid Inter. Sci.* **147**(148), 56–66.

Chang, J.-S., Hwang, Y. K., Jhung, S. H., Hong, D-Y. and Seo, Y.-K. (2007) Porous organic-inorganic hybrid materials and adsorbent comprising the same, USPTO Application: 20090263621 dated Aug. 1, 2007.

Chen, X., Lam, K. F., Zhang, Q., Pan, B., Arruebo, M. and Yeung, K. L. (2009) Synthesis of highly selective magnetic mesoporous adsorbent. *J. Phys. Chem. C* **113**(22), 9804–9813.

Chitrakar, R., Kanoh, H., Miyai, Y. and Ooi, K. (2001) Recovery of lithium from seawater using manganese oxide adsorbent ($H_{1.6}Mn_{1.6}O_4$) derived from $Li_{1.6}Mn_{1.6}O_4$. *Ind. Eng. Chem. Res.* **40**(9), 2054–2058.
Chubar, N., Avramut, C., Behrends, T. and Van Cappellen, P. (2009) Long-term sorption of Mn^{2+} by viable and autoclaved *Shewanella putrefaciens:* FTIR, XAFS and SEM characterization of the precipitates synthesized by the (initially) live bacteria. ECASIA '09 European Conference on *Surface & Interface Analysis*, Antalya, Turkey, October 17–23. Abstract book, p. 12.
Chubar, N., Behrends, T. and Van Cappellen, P. (2008b) Biosorption of metals (Cu^{2+}, Zn^{2+}) and anions (F^-, H_2PO_4-) by viable and autoclaved cells of the gram-negative bacterium *Shewanella putrefaciens. Colloids and Surfaces B: Biointerfaces* **65**, 126–133.
Chubar, N. I., Kanibolotskiy, V. A., Strelko, V. V. and Shaposhnikova, T. O. (2008a) Adsorption of anions onto inorganic ion exchangers. In *Selective Sorption and Catalysis on Active Carbons and Inorganic Ion Exchangers*, Strelko, V. (ed). Press: Naukova Dumka, Kiev, Vol. 2.
Chubar, N. I., Kanibolotskiy, V. A., Strelko, V. V., Shaposhnikova, T. O., Milgrandt, V. G., Zhuravlev, I. Z., Gallios, G. G. and Samanidou, V. F. (2005b) Sorption of phosphate ions on the hydrous oxides. *Colloids and Surfaces: A* **255**(1–3), 55–63.
Chubar, N. I., Kouts, V. S., Kanibolotskiy, V. A. and Strelko, V. V. (2006) Adsorption of anions onto sol-gel generated hydrous oxides. In NATO ARW series book, *Viable Methods of Soil and Water Pollution Monitoring, Protection and Remediation: Development and Use*, (ed.) Twardowska, I. Kluwer Publisher, The Netherlands, **69**, 323–338.
Chubar, N. I., Kouts, V. S., Samanidou, V. F., Gallios, G. G., Kanibolotskiy, V. A. and Strelko, V. V. (2005a) Sorption of fluoride, bromide, bromate and chloride ions on the novel ion exchangers. *J. Colloid Interf. Sci.* **291**(1), 67–74.
Chubar, N. I., Machado, R., Neiva Correia, M. J. and Rodrigeus de Carvalho, J. M. (2003b) Biosorption of copper, zinc and nickel by grape-stalks and cork biomasses. In NATO ARW series book: *Role of Interfaces in Environmental Protection*, (ed. Barany, S.). Kluwer Publisher, The Netherlands, pp. 339–353.
Chubar, N. I., Neiva Correia, M. J. and Rodrigeus de Carvalho, J. M. (2003c) Cork biomass as biosorbent for copper, zinc and nickel. *Colloids and Surfaces: A* **23**(1–3), 57–66.
Chubar, N. I., Neiva Correia, M. J. and Rodrigeus de Carvalho, J. M. (2004) Heavy metals biosorption on cork biomass: effect of pretreatment. *Colloids and Surfaces: A* **238**(1–3), 51–58.
Chubar, N. I., Strelko, V. V., Rodrigeus de Carvalho, J. M. and Neiva Correia, M. J. (2003a) Cork biomass as sorbent for color metals. *Water Chemistry and Technology* **25**(1), 33–38.
Clarke, T. D. and Wai, C. M. (1998) Selective removal of cesium from acid solutions with immobilised copper ferrocyanide. *Analytic Chemistry* **70**(17), 3708–3711
Clearfield, A. (1982) *Inorganic Ion Exchange Materials*. CRC Press Inc., Boca Raton, Florida.
Cochrane, E. L., Lu, S., Gibb, S. W. and Villaescusa, I. (2006) A comparison of low-cost biosorbents and commercial sorbents for the removal of copper from aqueous media, *J. Hazard. Mater. B* **137**, 198–206.

Colella, C. and Gualtieri, A. F. (2007) Cronstedt's zeolite. *Microp. Mesop. Mater.* **105**, 213–221.

Conant, J. B. (1950) *The Overthrow of Phlogiston Theory: The Chemical Revolution of 1775–1789*. Harvard University Press, Cambridge.

Cornelissen, E. R., Moreau, N., Siegers, W. G., Abrahamse, A. J., Rietverld, L. C., Grefte, A., Dignum, M., Amy, G. and Wessels, L. P. (2008) Selection of anionic exchange resins for removal of natural organic matter (NOM) fractions. *Water Res.* **42**(1–2), 413–423.

Dabrowski, A. (2001) Adsorption – from theory to practice. *Advances in Colloid Interface Sci.* **93**(1–3), 135–224.

Danish Ministry of the Environment (2007) BEK 1449 from 11, App. 1b, 2007.

Davis, T. A., Volesky, B. and Mucci, A. (2003) A review of the biochemistry of heavy metal biosorption by brown algae (review). *Water Res.* **37**, 4311–4330.

Drozdnik, I .D. (1997) Properties of carbon sorbents from coals with various degrees of metamorphism. *Fuel and Energy Abstracts* **3**(1), 29.

Elshazly, A .H. and Konsowa, A. H. (2003) Removal of nickel ions from wastewater using a cation-exchange resin in a batch-stirred tank reactor. *Desalination* **158**(1–3), 189–193.

Environmental Protection Agency (EPA) (2000) *National Primary Drinking Water Regulations*, http://www.epa.gov/safewater/sdwa/current_regs.html

Erdem-Şenatalar, A. and Tatlıer, M. (2000) Effects of fractality on the accessible surface area values of zeolite adsorbents. *Chaos, Solutions & Fractals* **11**(6), 953–960.

Ferguson, J. F. and Gavis, J. (1972) A review of the arsenic cycle in natural waters. *Water Res*. **6**, 1259–1274.

Fettig, J. (1999) Removal of humic substances by adsorption/ion exchange. *Water Sci. Technol.* **40**(9), 173–182.

Figuiredo, J. L., Pereira, M. F. R., Treitas, M. M. A. and Orfao, J. J. M. (1999) Modification of the surface chemistry of activated carbons. *Carbon* **37**, 1379–1389.

Fiore, S., Cavalcante, F. and Belviso, C. (2009) Patent application title: *Synthesis of Zeolites from Fly Ash*. Patent application number: 20090257948.

Fletcher, A. J., Kennedy, M. J., Zhao, X. B., Bell, J. B. and Mark Thomas, K. (2008) Adsorption of organic vapour pollutants on activated carbon. In NATO Science for Peace and Security Series C, *Environmental Security: Recent Advances in Adsorption Processes for Environmental Protection and Security*. Springer Netherlands Press, 29–54.

Franzreb, M., Hoell, W. H. and Eberle, S. H. (1995) Liquid-phase mass transfer in multicomponent ion exchange. 2. Systems with irreversible chemical reactions in the film. *Ind. Eng. Chem. Res.* **34**(8), 2670–2675.

Frost, R., Musumeci, A. W., Bostrom, T., Adebajo, M. O., Weier, M. L. and Martens, W. (2005) Thermal decomposition of hydrotalcite with chromate, molybdate or sulfate in the interlayer. *Thermochim. Acta* **429**, 179–187.

Górka, A., Bochenek, B., Warchoł, I., Kaczmarski, K. and Antos, D. (2008) Ion exchange kinetics in removal of small ions. Effect of salt concentration on inter- and intraparticle diffusion. *Chem. Eng. Sci.* **63**(3), 637–650.

Gregg, S. J. and Sing, S. W. (1982) Absorption, surface area and porosity, 2nd Ed., Academic press, New York.

Gun'ko, V. M. and Mikhalovky, S. V. (2004) Evaluation of slitlike porocity of carbon adsorbents. *Carbon* **42**, 843–849.

Gun'ko, V. M., Turov, V. V., Skubiszewska-Zieba, J., Leboda, R., Tsarko, M. D. and Palijczuk, D. (2003) *Ap. Surf. Sci.* **214**, 178–189.

Gunter, D. and Werner, S. (1997) US patent 4132671. *Process for the preparation of carbon black pellets*. http://www.freepatentsonline.com/4132671.html.

Gupta, S. K. and Chen, K. Y. (1978) Arsenic removal by adsorption. *J. Water Pollut. Contr. Fed.* **50**(3), 493–506.

Hall, A. and Stamatakis, M. G. (2000) Hydrotalcite and an amorphous clay minerals in high-magnesium mudstones from the Kozani basin, Greece. *J. of Sedimentary Researcher* **70**(3), 549–558.

Hathaway, S. W. and Rubel, F. (1987) Removing arsenic from drinking water. *J. Am. Water Works Assoc.* **79**(8) 61–65.

Helfferich, F. (1962) *Ion Exchange*. McGraw Hill, New York.

Henmi, T. and Sakagami, E. *Method of Producing Artificial Zeolite. European Patent EP0963949*. Filling date: 06/11/1999. Publication Date: 04/14/2004.

Henmi, T., Nakamura, T., Ubukata, T., Matsuda, H. and Tada, S. (2009) *Method of manufacturing artificial zeolite*. IPC8 Class: AC01B3904FI, USPC Class: 423703.

Hennessey, S. M., Friend, J., Elander, R. T. and Tucker, M. P. (2009) *Biomass pretreatment*, US Government Patent application N 20090053770. http://www.freshpatents.com/-dt20090226ptan20090053770.php

Hideki, K., Shigeki, K., Liang, R. and Atsushi, U, (2000) Manganese Oxide(Mn_2O_3) as adsorbent for cadmium. *Journal of Japan Society on Water Environment* **23**(2) 116–121.

Hoell, W. H. and Kalinichev, A. (2004) The theory of formation of surface complexes and its application to the description of multicomponent dynamic sorption systems. *Russ. Chem. Rev.* **73**, 351–370.

Hoell, W. H., Zhao, X. and He, S. (2002) Elimination of trace heavy metals from drinking water by means of weakly basic anion exchangers. *J. Water SRT – Aqua* **51**, 165–172.

Holtzapple, M. T., Lundeen, J. E., Sturgiss, R., Lewis, J. E. and Dale, B. E. (1992) Pretreatment of lignocellulosic municipal solid waste by ammonia fiber explosion. *Appl. Biochem. Biotechnology* **34**, 5–21.

Huang, C.-T. and Wu, G. (1999) Improvement of Cs leaching resistance of solidified radwastes with copper ferrocyanide (CFC)-vermiculite. *Waste Man.* **19**(4), 263–268.

Id., *Science de l'air: Studi su Felice Fontana*, Brenner: Cosenza, 1991.

Irving M. Abrams and John R. Millar (1997) "A history of the origin and development of macroporous ion-exchange resins", Reactive & Functional Polymers, **35**, 7–22.

IUPAC Recommendations 2003 (2004) *Pure Appl. Chem.*, **76**(4), 889–906.

Jain, A. K., Agrawal, S. and Singh, R. P. (1980) Selective cation exchange separation of secium(I) on chromium ferricyanide gel. *Anal. Chem.* **52**, 1364–1366.

Kang, S.-K., Choo, K-H. and Lim, K-H. (2003) Use of iron oxide particles as adsorbents to enhance phosphorus removal from secondary wastewater effluent. *Sep. Sci. Technol.* **38**(15), 3853–3874.

Karcher, S., Kornmüller, A. and Jekel, M. (2002) Anion exchange resins for removal of reactive dyes from textile wastewaters. *Water Res.* **36**(19), 4717–4724.

Katsoyiannis, I. A., Zouboulis, A. I. and Jekel, M. (2004) Kinetics of bacterial As(III) oxidation and subsequent As(V) removal by sorption onto biogenic oxides during groundwater treatment. *Ind. Eng. Chem. Res.* **43**(2), 486–493.

Kawamura, S., Kurotaki, K. and Izawa, M. (1969) Preparation and ion-exchange behavior of potassium ferrocyanide. *Bulletin of the Chemical Society of Japan* **42**, 3003–3004.

Kiefer, S. and Robens, E. (2008) Some of intriguing items in the history of volumetic and gravimetric adsorption measurements. *Journal of Thermal Analysis and Calorimetry* **94**(3), 613–618.

Kim, B. K., Kim, S.-H. and Alam, T. Kamaluddin (2000) Interaction of 2-amino, 3-amino and 4-aminopyridines with nickel and cobalt ferrocyanides. *Engineering Aspects* **162**(1), 89–97.

Kim, Y., Kim, C., Choi, I., Rengaraj, S. and Yi, J (2004) Arsenic removal using mesoporous alumina prepared via a templating method. *Environ. Sci. Technol.* **38** (3), 924–931.

Kim, Yu-H. (2000) Adsorption characteristics of cobalt on ZrO_2 and Al_2O_3 adsorbents in high-temperature water. *Sep. Sci. Technol.* **35**(14), 2327–2341.

Kogtev, L., Park, J. K., Pyo, J. K. and Mo, Y. K. (1998) *Biosorbent for heavy metals prepared from biomass*, United States Patent 5789204.

Kononova, O. N., Kholmogorov, A. G., Lukianov, A. N., Kachin, S. V., Pashkov, G. L. and Kononov, Y. S. (2001) Sorption of Zn(II), Cu(II), Fe(III) on carbon adsorbents from manganese sulfate solutions. *Carbon* **39**, 383–387.

Kumar, P., Barrett, M. B., Delwiche, M. J. and Stroeve, P. (2009) Methods for pretreatment of Lignocellulosic Biomass for efficient hydrolysis and biofuel production. *Ind. Eng. Chem. Res.* **48**(8), 3713–3729.

Kunin, R. (1982) *Ion Exchange Resins*. Robert E. Krieger Publishing, Company, Melbourne, FL, pp. 3 and 130.

Lach, J., Okoniewska, E., Neczaj, E. and Kacprzak, M. (2007) Removal of Cr(III) cations and Cr(VI) anions on activated carbons oxidized by CO2. *Desalination* **206**, 259–269.

Lefevre, G. (2004) In situ Fourier-transfrom infrared spectroscopy studies of inorganic ions adsorption on metal oxides and hydroxides. *Adv. Colloid Interf. Sci.* **107**, 109–123.

Likholobov, V. A. (2007) Institute of Hydrocarbon Processing, Siberian Branch, Russian Academy of Sciences advances of science and practice in solving problems of chemical hydrocarbon processing. *Russian Journal of General Chemistry* **12**(17), 1070–3632.

Liu, Z., Wei, Y., Qi, Y., Liu, X., Zhao, Y. and Liu Z. (2006) Synthesis of ordered mesoporous Zr-P-Al materials with high thermal stability. *Microp. Mesop. Mater.* **91**, 225–232.

Long, C., Lu, J. D., Li, A., Hu, D., Liu, F. and Zhang, Q. (2008) Adsorption of naphthalene onto the carbon adsorbent from waste ion exchange resin: Equilibrium and kinetic characteristics. *J. Hazard. Mater.* **150**, 656–661.

Lopes, T., Ramos, B. E., Gomes, R., Novaro, O., Acosta, D. and Figueras, F. (1996) Synthesis and characterisation of sol-gel hydrotalcites, structure and texture. *Langmuir* **12**, 189–192.

Loureiro, J. M. and Kartel M. T. (2006) *Combined and Hybrid Adsorbents: Fundamentals and Application*. Springer-Verlag New York Inc.

Lucy, C. A. (2003) Evolution of ion-exchange: From Moses to the Manhattan project to modern times. *J. Chromatography A* **1000**(1–2), 711–724.

Malikov, I. N., Noskova, Yu, A., Karaseva, M. S. and Perederii, M. A. (2007) Granulated sorbents from wood wastes. *Solid Fuel Chemistry* **41**(2), 100–106.

Manceau, A., Drits, V. A., Silvester, E., Bartoli, C. and Lanson, B. (1997) Structural mechanism of Co^{2+} oxidation by the phyllomanganate buserite. *Am. Mineral.* **82**, 1150–1175.

Mandich, N. V., Lalvani, S. B., Wiltkowski, T. and Lalvani, L. S. (1998) Selective removal of chromate anion by a new carbon adsorbent. *Metal Finishing* **96**(5), 39–44.

Manjare, S. D., Sadique, M. H. and Ghoshal, A. K. (2005) Equilibrium and kinetics studies for As(III) adsorption on activated alumina and activated carbon. *Environ. Technol.* **26**(12), 1403–1410.

Manning, B. A., Fendorf, S. E., Bostick, B. and Suarez, D. L. (2002) Arsenic(III) oxidation and arsenic(V) adsorption reactions on synthetic birnessite. *Environ Sci. Technol.* **36**(5), 976–981.

Manos, Manolis J., Nan Ding and Kanatzidis, Mercouri, G. (2008) Layered metal sulfides: Exceptionally selective agents for radioactive strontium removal. *PNAS* **5**(10), 3696-3699.

Mansoor, A. and Moradi, S. I. (2009) Removal of naphthalene from petrochemical wastewater streams using carbon nanoporous adsorbent. *Ap. Surf. Sci.* **255**, 5041–5047.

Margadonna, S. (2008) *Fullerene-Related Materials: Recent Advances in Their Chemistry and Physics*, 1st edition, 700 pp, Springer.

Marshall, W. E. and Wartelle, L. H. (2004) An *anion exchange* resin from soy-bean hulls. *J. Technol. Biotechnol.* **79**, 1286–1292.

Matulionytė, J., Vengris, T., Ragauskas, R. and Padarauskas, A. (2007) Removal of various components from fixing rinse water by anion-exchange resins. *Desalination*, **208**, 81–88.

Meleshevych, I., Pakhovchyshyn, S., Kanibolotsky, V. and Strelko, V. (2007) Rheological properties of hydrated zirconium dioxide. *Colloids and Surfaces A* **298**, 274–279.

Melián-Cabrera, I., Kapteijn, F. and Moulijn, J. A. (2005) Innovations in the synthesis of Fe-(exchanged)-zeolites. *Catal. Today* **110**, 255–263.

Miers, J. A. (1995) Regulation of ion exchange resins for the food, water and beverage industries. *Reactive Polymers* **24**, 99–107.

Misra, C. and Genito, J. R. (1993) US Patent 5270278 – *Alumina coated with a layer of carbon as an absorbent*. Issued Dec. 14, 1993.

Moore, J. N., Walker, J. R. and Hayes, T. H. (1990) Reaction scheme for the oxidation of As(III) to arsenic(V) by birnessite. *Clays Clay Miner*. **38**, 549–555.

Mrowiec-Białoń, J., Jarzbski, A. J., Lachowski, A. I., Malinowski, J. J. and Aristov Y. I. (1997) Effective inorganic hybrid adsorbents of water vapor by the sol-gel method. *Chem. Mater.* **9**(11), 2486–2490.

Mui, E. L. K., Ko, D. C. K. and McKay, G. (2004) Production of active carbons from waste tyres – a review. *Carbon* **42**, 2789–2805.

Namasivayam, C., Sangeetha, D. and Gunasekaran, R. (2007) Removal of anions, heavy metals, organics and dyes from water by adsorption onto a new activated carbon from Jatropha Husk, in agro-industrial solid waste. *Process Safety and Environmental Protection* **85**(2), 181–184.

National Health and Medical Research Centre (1996) *Australian drinking water guidelines – Summary, Australian Water and Wastewater Association*, Artamon.

Natural Recourses Defense Council (2000) Arsenic and old laws: A scientific and public health analysis of arsenic occurrence in drinking water, its health effects, and EPA's outdated arsenic tap water standard, available at: ww.nrdc.org/water/drinking/arsenic/aolinx.asp.

New Jersey Department of Environmental Protection (2004) Safe drinking water act regulations N.J.A.C., **7**(10), 1–83.

Novoselova, L. Y. and Sirotkina, E. E. (2008) Peat-based carbons for purification of the contaminated environments. *Solid Fuel Chemistry* **42**(4), 251–262.

Ooi, K., Miyai, Y. and Katoh, S. (1986) Recovery of. lithium from seawater by manganese oxide adsorbent. *Sep. Sci. Technol.* **21**(8), 755–766.

Ouvrard, S., Simonnot, M. O. and Sardin, M. (2002a) Reactive behavior of natural manganese oxides toward the adsorption of phosphate and arsenate. *Ind. Eng. Chem. Res.* **41**, 2785–2791.

Ouvrard, S., Simonnot, M. O., Donato, P. and Sardin, M. (2002b) Diffusion-controlled adsorption of arsenate on a natural manganese oxide. *Ind. Eng. Chem. Res.* **41**, 6194–6199.

Palmer, S. and Frost, R. L. (2009) The effect of synthesis temperature on the formation of hydrotalcites in bayer liquor: A vibration spectroscopic analysis. *Applied Spectroscopy* **63**(7), 748–752.

Park, H. G., Kim, T. W., Chae, M. Y. and Yoo, I.-K. (2007) Activated carbon-containing alginate adsorbent for the simultaneous removal of heavy metals and toxic organics. *Process Biochem.* **14**(10), 1371–1377.

Patzak, M., Dostalek, P., Fogarty, R. V., Safarik, I. and Tobin, J. M. (2004) Development of magnetic biosorbents for metal uptake. *Biotechnol. Techniques* **11**(7), 483–487.

Petrus, R. and Warchol, I. (2005) Heavy metal removal by clinoptilolite. An equilibrium study in multi-component systems. *Water Res.* **39**(5), 819–830.

Pokonova, Y. V. (2001) Carbon adsorbents from coal pitch. *Chemistry and Technology of Fuels* **37**(3), 207–211.

Psareva, T. S., Zakutevskyy, O., Chubar, N.I., Strelko V. V. Shaposhnikova, T. O., Rodriges de Carvalho, J. M. and Neiva Correia, J. M. (2005) Uranium sorption on cork biomass. *Colloid and Surfaces: A* **252**(2–3), 231–236.

Quirarte-Escalante C. A., Soto, V., De La Cruz, W., Porras, G. R., Rangel, G., Manriques, R. and Gomez-Salazar, S. (2009) Synthesis of hybrid adsorbents combining sol-gel processing and molecular imprinting applied to lead removal from aqueous streams. *Chemistry of Materials* **21**(8), 1439–1450.

Radionuclides (2004) Final Rule, 40 CFR Parts 9, 141 and 142.

Roberts, L. C., Hug, S. J., Ruettimann, T., Khan, A. W. and Rahman, M. T. (2004) Arsenic removal with iron(II) and iron(III) in waters with high silicate and phosphate concentrations. *Environ. Sci. Technol.* **38**, 307–315.

Roddick-Lanzilotta, A. J., McQuillan, A. J. and Craw, D. (2002) Infrared spectroscopic characterization of arsenate(V) ion adsorption from mine waters, Macraes mine, New Zealand. *Appl. Geochem.* **17**, 445–454.

Rodrigues-Reinoso, F. and Molina-Sabio, M. (1998) Textural and chemical characterization of microporous carbons. *Adv. Colloid Interface Sci.* **76–77**, 271–294.

Saha, B., Bains, R. and Greenwood, F. (2005) Physicochemical characterization of granular ferric hydroxide (GFH) for arsenic(V) sorption from water. *Sep. Sci. Technol.* **40**(14), 2909–2932.

Schiewer, S. and Patil, P. B. (2008) Pectin-rich fruit wastes as biosorbents for heavy metal removal: Equilibrium and kinetics. *Biores. Technol.* **99**(6), 1896–1903.

Sengupta, S. and SenGupta, A. K. (1997) Heavy-metal separation from sludge using chelating ion exchangers with nontraditional morphology. *Reactive & Functional*

Polymers **35**, 111–134.
Seniavin, M. M. (1981) *Ion Exchange*. Nauka Publishing, Moscow. (In Russian).
Sharygin, L., Muromskiy, A., Kalyagina, M. (2007) A granular inorganic cation-exchanger selective to cesium. *Journal of Nuclear Science and Technology* **44**(5), 767–773.
Shectman, J. (2003), *Groundbreaking Scientific Experiments, Inventions, and Discoveries of the 18th Century*. Connecticut: Greenwood Press.
Shen, W., Zhijie, L. and Liu, Y. (2008) Surface chemical functional groups modification of porous carbon. *Recent Patents on Chemical Engineering* **1**, 27–40.
Singh, T. S. and Pant, K. K. (2004) Equilibrium, kinetics and thermodynamic studies for adsorption of As(III) on activated alumina. *Sep. Purif. Technol.* **36**, 139–147.
Sparks, D. E., Morgan, T., Patterson, P. M., Adam, T., Morris, E. and Crocker, M. (2008) New sulfur adsorbents derived from layered double hydroxides I: Synthesis and COS adsorption. *Appl.Catal. B- Environ.* **82**(3–4), 190–198.
Tanaka, Y. and Tsuji, M. (1994) New synthetic method of producing α-manganese oxide for potassium selective adsorbent. *Materials Research Bulletin* **29**(11), 1183–1191.
Tananaev, I. B., Seifer, G. B., Kharitonov, Y. Y., Kuznetsov, B. G. and Korolkov A. P. (eds) (1971) *Chemistry of Ferrocyanides*Nauka press, Moscow.
Thirunavukkarasu, O. S., Viraghavan, T. and Suramanian, K.S. (2003) Arsenic removal from drinking water using iron-oxide coated sand. *Water Air Soil Pollut.* **142**, 95–111.
Thompson, H. S. (1850) Absorbent power of soils. *J. R. Agric. Soc. Engl.* **11**, 68.
Tomlinson, A. A. G. (1998) Modern Zeolites. Press: Trans Tech Publications Inc. Laubisrutistr, Switzerland.
Tomoyuki, K., Kosuke, A., Toshio S. and Yoshio O. (2007) Synthesis and characterization of Si-Fe-Mg mixed hydrous oxides as harmful ions removal materials. *J Soc Inorg Mater Jpn* **14**(327), 1345–3769.
Treacy, M. M. J. and Higgins, J. B. (2007) *Collection of Simulated XRD Powder Diffraction Patterns for Zeolites, 5th revised edition*. Elsevier, Amsterdam.
Kim, S. J., Park, Y. Q. and Moon, H. (2007) Removal of copper ions by a cation-exchange resin in a semifluidized bed. *Korean Journal of Chemical Engineering* **15**(4), 417–422.
United States Patent 5865898, http://www.freepatentsonline.com/5865898.html.
USEPA (1999) Technologies and Costs for Removal of Arsenic from Drinking Water, Draft Report, EPA-815-R-00-012, Washington, DC.
Valentine, R. L., Mulholland, T. S. and Splinter, R. C. (1992) *Radium removal using sorption to filter preformed hydrous manganese oxides*. Report for the American Water Works. Association Research Foundation.
Valinurova, E. R., Kadyrova, A. D., Sharafieva, L. R. and Kudasheva, F. Kh. (2008) Use of activated carbon materials for wastewater treatment to remove Ni(II), Co(II), and Cu(II) ions. *Russian Journal of Applied Chemistry* **81**(11), 1939–1941.
Venkatesan, K. A., SathiSasidharan, N. and Wattal, P. K. (1997) Sorption of radioactive strontium on a silica-titania mixed hydrous oxide gel. *Journ. of Analytical and Nuclear Chemistry* **20**(1), 55–58.
Volesky, B. (1990a) Biosorption by fungal biomass. In *Biosorption of Heavy Metals*. Volesky, B. (ed) Florida: CRC press, pp. 139–171.
Volesky, B. (1990b) Introduction. In *Biosorption of Heavy Metals*. Volesky, B. (ed) Florida: CRC press, pp. 3–5.
Volesky, B. (1990c) Removal and recovery of heavy metals by biosorption. In *Biosorption of Heavy Metals*. Volesky, B. (ed), Florida: CRC press, pp. 8–43.

Volesky, B. (1994) Advances in biosorption of metals — selection of biomass types. *FEMS, Microbiol Rev.* **14**, 291–302.

Volesky, B. (2001) Detoxification of metal-bearing effluents: biosorption for the next century. *Hydrometallurgy* **59**, 203–216.

Volesky, B. (2003) Biosorption process simulation tools. *Hydrometallurgy* **71**, 179–90.

Volesky, B. (2007) Biosorption and me. *Water Res.* **41**, 4017–29.

Volesky, B. and Holan, Z. R. (1995) Biosorption of heavy metals. *Biotechnol Prog.* **11**, 235–50.

Volesky, B. and Naja, G. (2005) Biosorption: application strategies. *16th Internat. Biotechnol. Symp. Compress Co., Cape Town, South Africa.*

Volesky, B, May, H. and Holan, Z. R. (1993) Cadmium biosorption by Saccharomyces cerevisiae. *Biotechnol Bioeng.* **41**, 826–829.

Volgin, V. V. (1979) News of the Academy of Sciences. *Inorganic Materials* **15**, 1084–1089.

Vollmer, D. L. and Gross, M. L. (2005) Cation-exchange resins for removal of alkali metal cations from oligonucleotide samples for fast atom bombardment mass spectrometry. *J. Mass Spectrometer* **30**(1), 113–118.

Voyutsky, S. S. (1964) *Colloid Chemistry.* Chemistry Press, Moscow.

Walsh, R. (2008) *Development of a biosorption column utilizing seaweed based biosorbents for the removal of metals from industrial waste streams.* PhD thesis, Waterford Institute of Technology: http://repository.wit.ie/1031/.

Walt, D. K. (2003) *Applied Biochemistry and Biotechnology*, Humana Press, **10**(1–3), Spring 2003.

Wang, J. and Chen, C. (2009) Biosorbents for heavy metals removal and their future. *Biotechnol. Adv.* **27**(2), 195–226.

Wang, J. S. and Wai, C. M. (2004) Arsenic in drinking water—a global environmental problem. *J. Chem. Educ.* **81**, 207–213.

Wang, L. K., Hung, Y-T., Lo, H. H. and Yapijakis, C. (eds) (2006) *Waste Treatment in the Process Industries*. Taylor & Francis Group, New York.

Watanabe, S. Velu, S. Ma, X. and Song, C. S. (2003) Preprint Paper – American Chemical Society, *Division Fuel Chemistry* **48**(2), 695–696.

Wayne, E. M. and Wartelle, L. (2004) An anion exchange resin from soybean hulls. *J. of Chemical Technol. Biotechnol.* **79**(11), 1286–1292.

Whitehead, P. (2007) Medicine from animal cell culture. In *Water Purity and Regulations*, 696 pp. Glyn Stacey and John Davis (eds), John Wiley & Sons.

Wilkie, J. A. and Hering, J. G. (1996) Adsorption of arsenic onto hydrous ferric oxide: Effects of adsorbate/adsorbent ratios and co-occurring solutes. *Colloid Surf. A: Physicochem. Eng. Aspects* **107**, 97–110.

Yamamoto T., Taniguchi A., Dev S., Kubota E., Osakada K. and Kubota K. (1990) New organosols of nickel sulfides, palladium sulfides, manganese sulfide, and mixed metal sulfides and their use in preparation of semiconducting polymer-metal sulfide composites. *Colloid Polymer Science* **269**(10), 969–971.

Yin, C. Y., Arpna, M. K. and Daud, W. M. A. W. (2007) Review of modification of activated carbon for enhancing contaminant uptake from aqueous solutions. *Sep. Purif.Ttechnol.* **52**, 403–415.

Zhang, W., Zou, L.' and Wang, L. (2009) Photocatalytic TiO_2/adsorbent nanocomposites prepared via wet chemical impregnation for wastewater treatment: A review. *Appl. Catal A-General* **371**(1–2), 1–9.

Zhang, Y., Yang, H., Zhou, K. and Ping, Z. (2007) Synthesis of an affinity adsorbent based on silica gel and its application in endotoxin removal. *Reactive & Functional Polymers* **67**, 728–736.

Zhuravlev, I., Kanibolotsky, V., Strelko, V., Gallios, G. and Strelko, V. Jr. (2004) Novel high porous spherically granulated ferrophosphatesilicate gels. *Materials Research Bulletin* **39**(4–5), 737–744.

Zhuravlev, I. Z. (2005) *Sol-gel synthesis and properties of the ion exchangers based on composite phosphates of polyvalent metals and silica*, Ph.D. Thesis (Chemistry), Kiev, Ukraine.

Zhuravlev, I. Z. and Strelko V. V. (2006) Template effect of the M^{3+}-cations in the course of the synthesis of high dispersed titanium and zirconium phosphate. In *Combined and Hybrid Adsorbent, NATO Security through Science Series*, Springer, Netherlands. pp. 93–98.

Zhuravlev, I., Zakutevsky, O., Psareva, T., Kanibolotsky, V., Strelko, V., Taffet, M. and Gallios, G. (2002) Uranium sorption on amorphous titanium and zirconium phosphates modified by Al^{3+} or Fe^{3+} ions. *Journal of Radioanalytical and Nuclear Chemistry* **254**(1), 85–89.

Zolotov, Yu. A. (1998), Analytical Chemistry in Russia. *Fresenius J. Anal. Chem.* **361**, 223–226.

6

Membrane Processes and Coagulation for Micropollutant Treatment

O. Lefebvre, Lai Yoke Lee and How Yong Ng

6.1 COAGULATION

Coagulation is one of the oldest processes commonly used for removal of suspended solids, while enhanced coagulation using inorganic or polymeric coagulants has been applied for micropollutant removal in water and wastewater treatment. Enhanced coagulation is a popular option for upgrading existing treatment facility to incorporate removal of selected micropollutants as it offers several advantages especially in terms of cost effectiveness, simplicity in design, and ease of operation. However, the major drawback in enhanced coagulation for micropollutant removal is the low to insignificant effect on some of the micropollutants in drinking water and wastewater treatment.

Improvement in the coagulation process for micropollutant removal has been reported using oxidation-coagulation/precipitation and coagulation-membrane separation (Bodzek and Dudziak, 2006; Lee *et al.*, 2009; Lim and Kim, 2009). Pharmaceutical compounds such as diclofenac using Fe (VI) oxidation-coagulation attained more than 95% removal at 5 mg Fe/L (Lee *et al.*, 2009) as compared with only more than 65% with enhanced coagulation using 50 mg/L $FeCl_3$ or $Al_2(SO_4)_3$ (Carballa *et al.*, 2005). Integrating coagulation-NF system was able to provide at least 18.5% increase in the removal of estrogen as compared with using NF system alone (Bodzek and Dudziak, 2006). Applications of enhanced coagulation and oxidation-coagulation for micropollutant removal, and mechanisms and controlling factors in these processes will be presented in the following sections.

6.1.1 Enhanced coagulation

In drinking water treatment using conventional filtration treatment, improved disinfection by-products (DBP) removal is achieved by applying enhanced coagulation. Enhanced coagulation in drinking water treatment is defined as the addition of excess coagulant and possibly accompanied by reduced coagulation pH (Edwards *et al.*, 2003). This has been introduced as a requirement in the Environmental Protection Agency (EPA) DBP Rule (Freese *et al.*, 2001).

A cost comparison between the advanced water treatment processes such as ozonation and granular activated carbon (GAC) with enhanced coagulation concluded that the latter is more cost-effective for smaller treatment works with capacity less than 175 million litres per day and for cleaner influent quality (i.e., with TOC concentration less than 5 mg/L) (Freese *et al.*, 2001). The cost of treatment of the clean influent water (TOC <5 mg/) using enhanced coagulation was at least 25% lower for a 175 million litres per day plant as compared to a plant with double of this capacity. Table 6.1 summarizes the removals of some of the micropollutants using enhanced coagulation method.

Sorption of micropollutants onto coagulant surface is the main removal mechanism in enhanced coagulation. The solid-liquid partitioning coefficient (K_d), which is the concentration ratio between solid and liquid phase at equilibrium conditions (Hemond and Fechner-Levy, 2000), can be used to determine the sorption behaviour of micropollutants. Two interactions between the micropollutants and the coagulants are involved; namely lipophilic interactions which are indicated by the octanol-water (K_{ow}) and organic carbon (K_{oc}) partition coefficients, and electrostatic interactions which are associated with the dissociation constant (pK_a) of the pollutants and the weak Van der Waals bonds with the coagulants (Stumm and Morgan, 1996). Therefore, the physical-chemical properties of the micropollutants and how they change with the aqueous environment are important factors that influence the micropollutants removal. Parameters that influence the physical-chemical properties of micropollutants in the aqueous environment include pH, alkalinity, choice of coagulants and their optimum concentrations (McGhee, 1991).

Table 6.1 Micropollutant removal using enhanced coagulation methods

Micropollutant	log K_{ow}	pK_a	Coagulant	Dosage	Removal Efficiency (%)
Anthropogenic					
Trihalomethane formation potential (THMFP)[1]	–	–	Iron (III) chloride	Up to 30 mg/L	Up to 40%
Fragrances					
Galaxolide[2]	5.9–6.3	–	Aluminum polychloride	17.5% w/w	63
Tonalide[2]	4.6–6.4	–	Aluminum polychloride	17.5% w/w	71
Pharmaceutical					
Diazepam[2]	2.5–3.0	3.3–3.4	Iron (III) chloride	50 mg/L	~25
Naproxen[2]	3.2	4.2	Iron (III) chloride	50 mg/L	~ 20
Diclofenac[2]	4.5–4.8	2.5–3.0	Iron (III) chloride	50 mg/L	>65
			Aluminum sulfate	50 mg/L	>65
Endocrine disruptors					
Bisphenol A (BPA)[3]		10.2[5]	Iron (III) chloride	Up to 200 mg/L	Up to 20
Diethylhexylphthalate (DEHP)[3]			Iron (III) chloride	Up to 200 mg/L	Up to 70
17β-Estradiol (E2)[4]	4.01	10.4[5]	Iron (III) sulfate	12.2 mg/L	15
			Polyaluminum chloride	5.4 mg/L	15

(continued)

Table 6.1 (*continued*)

Micropollutant	log K_{ow}	pK_a	Coagulant	Dosage	Removal Efficiency (%)
Estriol (E3)[4]	2.45	–	Iron (III) sulfate	12.2 mg/L	20
			Polyaluminum chloride	5.4 mg/L	30
Diethylstilbestrol (DES)[4]	5.07	–	Iron (III) sulfate	12.2 mg/L	25
			Polyaluminum chloride	5.4 mg/L	40

[1] Freese *et al.* (2001)
[2] Carballa *et al.* (2005)
[3] Asakura and Matsuto (2009)
[4] Bodzek and Dudziak (2006)
[5] Deborde *et al.* (2005)

a) Effects of physical-chemical properties of micropollutants

Micropollutants with high K_{ow} values are less soluble in water and would have a higher tendency to sorb onto particles (Hemond and Fechner-Levy, 2000). Hence, micropollutants with higher K_{ow} would have a higher tendency to be removed from the liquid phase. This could be observed from the results reported by Carballa *et al.* (2005) that with different coagulants, diclofenac (log K_{ow} of 4.5–4.8) experienced removal efficiencies of 50–70% as compared with naproxen (log K_{ow} of 3.2) which only achieved removal efficiencies in the range of 5–20% (Table 6.1).

Similarly, synthetic estrogens with higher K_{ow} values such as mestranol (log K_{ow} of 4.67) had about 40% higher sorption to sediment as compared with estriol (log K_{ow} of 2.81) (Table 6.2) (Lai *et al.*, 2000). Competitive effects were also observed between the micropollutants for binding sites on the coagulants or sediments. Lai *et al.* (2000) reported that the sorption of other estrogens such as estriol and mestranol was suppressed by 89 and 31%, respectively, with the addition of a superhydrophobic synthetic estrogen, estradiol valerate (with higher log K_{ow} of 6.41). The suppression effect of this competitive binding was more significant for compounds with lower hydrophobicity as noted from the higher suppression experienced by estriol. Enhanced coagulation was also determined to be more effective for the removal of high molecular weight DBP compounds (> 30 kDa). At pH 7.5, adsorption was reported to be the main contributor for DBP compounds ($>$30 kDa) removal by the coagulants (Zhao *et al.*, 2008). However, in the treatment of natural water and wastewater when a complex mixture of micropollutants is present, the effect of enhanced coagulation towards removal of micropollutants would differ and further study is required.

The sorption kinetics of the synthetic estrogen on sediments showed rapid sorption within the initial 0.5 hr followed by a slower sorption rate which subsequently achieved a steady decrease (Lai *et al.*, 2000). The estrogen removal rate in a batch experiment followed a 3-stage sorption pattern. A general trend in the sorption experience in the batch test is shown in Figure 6.1. The initial stage 1 occurred at a rapid rate due to the availability of active binding sites. Upon gradual saturation of these sites, the rate was gradually reduced reaching close to a steady state. This could also be due to the lower availability of micropollutants in the aqueous phase. Subsequent desorption of micropollutants from the solids occurred, reducing the net amount of micropollutants sorbed by the solids. Hence, for an optimum removal through enhanced coagulation, a rapid mixing and gradual flocculation should fall within Stages 1 and 2 and removal of the flocs would be required before Stage 3 to avoid desorption of micropollutants back into the aqueous phase.

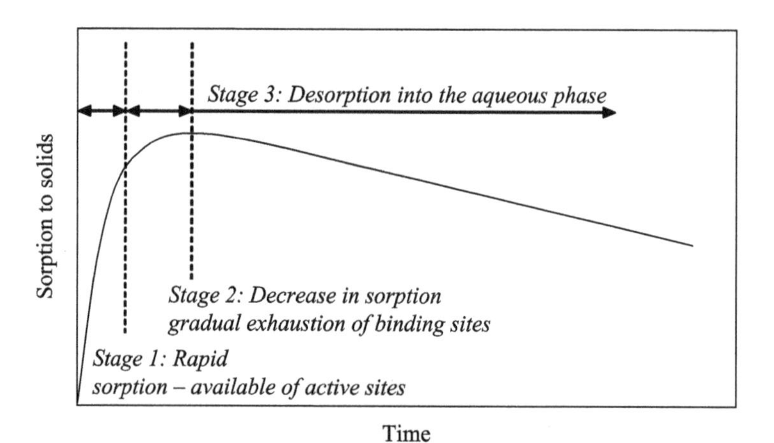

Figure 6.1 Typical sorption pattern of micropollutants onto solids

Similar to the effect of K_{ow} on the removal of micropollutants, synthetic estrogen with a higher K_{ow} was also reported to have a higher sorption rate onto the solids in Stage 1 and a lower desorption rate in Stage 3 (Lai *et al.*, 2000). A summary on the physical-chemical properties and the sorption constant of selected synthetic estrogens determined by Lai *et al.* (2000) is given in Table 6.2.

Table 6.2 Physical-chemical properties and sorption characteristics of some selected synthetic estrogens (extracted from Lai *et al.*, 2000)

Micropollutant	Molecular Weight	Water Solubility (mg/L)	log K_{ow}	Sorption Constant (1/n)	Sorption to Sediment (ng/g)
Estriol	288.39	13	2.81	0.57	3.2
Estradiol	272.39	13	3.94	0.67	4.1
Mestranol	310.42	0.3	4.67	0.78	5.5

b) Choice of coagulants and dosage

The coagulation-flocculation studies on removal of pharmaceutical and personal care products (PPCP) by Carballa *et al.* (2005) reported that the coagulant dose (using ferrous chloride ($FeCl_3$) of 250–350 mg/L, aluminum sulphate ($Al_2(SO_4)_3$) of 250–350 mg/L and aluminum polychloride (PAX) of 700–950 mg/L) and temperature (12 or 25°C) tested did not have a significant influence on the PPCPs removal. However, the removal efficiencies of specific PPCPs were influenced by the type of coagulants used. Overall, $FeCl_3$ (250 mg/L at 25°C) was able to provide more than 50% removal for galaxolide, tonalide and diclofenac, while about 20–25% removal for diazepam and naproxen. Table 6.3 summarises the highest removal efficiencies of different PPCPs achieved among the three coagulants tested at 25°C as reported by Carballa *et al.* (2005).

Table 6.3 Highest removal efficiencies of different PPCPs and the respective coagulants used (Carballa *et al.*, 2005)

Type of PPCP	Type of Coagulant (Dosage in mg/L)	Removal Efficiency (%)
Galaxolide	PAX (850)	63
Tonalide	PAX (850)	71
Diclofenac	$FeCl_3$ (250)	70
Diazepam	$FeCl_3$ (250)	25
Naproxen	$FeCl_3$ (250)	20

Zorita *et al.* (2009) reported that by applying 0.07 mg/L $FeCl_3$ in a tertiary treatment comprised of coagulation and flocculation for sewage wastewater treatment system, a removal efficiency of more than 55% was achieved for the antibiotics tested, namely, ofloxacin, norfloxacin and ciprofloxacin. It was proposed that the main removal mechanism was due to adsorption of these micropollutants onto flocs. The addition of $FeCl_3$ did not significantly enhance the removal of acidic group of pharmaceutical compounds, such as ibuprofen, naproxen, diclofenac and clofibric acid in the wastewater treatment (less than 25% removal efficiency for this group of compounds). This could be due to the low $FeCl_3$ dose used which led to insignificant removals of naproxen and diclofenac (Carballa *et al.*, 2005), whereas a higher dose of 250 mg/L of $FeCl_3$ used in the study by Zorita *et al.* (2009) induced removal efficiencies of 20 and 70% for naproxen and diclofenac, respectively. As these acidic compounds possess a physical-chemical property of pK_a 3–5, they would be partially ionized in aqueous phase. The effect of doubling the ferric chloride dose also provided an improvement in the removal of bisphenol A by enhanced coagulation from 5% to about 20% (Carbella *et al.*, 2005). The addition of coagulants would enhance the binding of these compounds onto the suspended solids for subsequent removal from the aqueous phase (Carbella *et al.*, 2005).

c) pH and alkalinity

Variation in pH could lead to the change in coagulants species and/or charge neutralization. Yan *et al.* (2008) and Zhao *et al.* (2008) demonstrated the changes in Al species with pH using ferron assay and electrospray ionization (ESI) mass spectrometry, respectively. Extensive researches have been performed by Yan *et al.* (2007, 2008) on the effect of pH/alkalinity on enhanced coagulation with PACls. The low pH in enhanced coagulation, however, is a major drawback in its applications which could lead to corrosion problems in plant infrastructures (Edwards *et al.*, 2003). PACl is a type of pre-hydrolyzed coagulant which can be applied to reduce the pH drop after coagulation. The different hydrolyzed Al species are summarized in Table 6.4 (Yan *et al.*, 2007).

Table 6.4 Hydrolyzed Al species (Yan *et al.*, 2007)

Hydrolyzed Group	Species	Molecular Weight (Da)
Al_a	Monomer – Al^{3+}, $Al(OH)^{2+}$, $Al(OH)_2^{+}$ Dimer – $Al_2(OH)_2^{4+}$ Trimer – $Al_3(OH)_4^{5+}$ Small polymers	<500
Al_b	Tridecamer – $Al_{13}O_4(OH)_{24}^{7+}$ (more commonly known as Al_{13})	500–3000
Al_c	Large polymer or colloidal species	>3000

The degree of hydrolysis is expressed using basicity (B) value which represents the ratio of hydroxide-to-aluminum ratio (Yan *et al.*, 2008). Generally higher B value corresponds to lower Al_a and higher Al_b portion. The degree of hydrolysis is significantly influenced by the pH. Table 6.5 summarizes the dominant Al species at different pH region. Al_b has a tendency to form Al_c with aging. Together with pH, coagulant dose also plays an important role in determining the stability of the species present. Wang *et al.* (2007) reported that at higher PACl dosage above 2 mol Al/L, Al_b was not detected. However, using a different coagulant, such as the nano-Al_{13} coagulant, showed that Al_b species were relatively stable at coagulant concentrations of 0.11–2.11 mol Al/L even up to 30 days of aging (Wang *et al.*, 2007). The presence of preformed Al_b in the pre-hydrolyzed PACl was noted to enhance natural organic matter (NOM) removal in the surface water (Yan *et al.*, 2008).

Table 6.5 Dominant Al species at different pH region for 3 different types of coagulants (Yan *et al.*, 2008)

Type of Coagulant	B	Dominant Al Species		
		$pH < 5.0$	$5.5 < pH < 7.5$	$pH > 9.0$
$AlCl_3$	0	Al_a	Al_b	Al_a
$PACl_l$	1.6	$Al_a \approx Al_b$	Al_b	$Al_a \approx Al_b$
$PACl_{20}$	2.0	Al_b	Al_b	Al_b

In addition to pH, the characteristics of the raw water also affect the performance of the coagulants. Yan *et al.* (2008) demonstrated that a higher coagulant dose was required to depress the pH of the Yellow River water, which had an alkalinity of 3.5 times higher than the water from the Pearl River, to a pH range (about pH 5.5–6.5 based on optimal NOM removal) favourable for coagulation.

Another controlling factor is the effect of pH on micropollutants. The change in reaction pH will affect dissociation of the micropollutants which is given by the dissociation constant, pK_a. The removal of DBP (such as Trihalomethane (THM) and Haloacetic acid (HAA)) precursors is affected by the pH of the reaction solution. Removal of THM and HAA precursors using PACl coagulation is mainly due to the nature of the THM and HAA precursors (Zhao *et al.*, 2008). THM precursors contain more aliphatic structure while HAAs precursors are mainly aromatic. Charge neutralisation precipitation is responsible for the removal of the negatively-charged aliphatic THM precursors which could be achieved at pH 5.5. Two possible effects could occur: neutralization of the negatively charged compounds by the monomeric

Al species, or neutralization by the hydrogen ions and Al ions (Yan *et al.*, 2007). At a higher pH between 5.5–7.5, the presence of both Al_{13} and $Al(OH)_3$ would have resulted in simultaneous removal of THM precursors by charge neutralization precipitation and adsorption. In the case of HAA precursors, self-aggregations of the aromatic and hydrophobic functional groups occurred under acidic conditions, while under alkaline conditions, absorption onto flocs and subsequent removal through sweep flocculation occur.

6.1.2 Coagulation-oxidation

Removal of micropollutants can be further enhanced with the used of more advanced coagulation process such as oxidation-coagulation. Oxidizing agents have the potential of achieving both oxidation and coagulation, such as ferrate (Fe)(VI) for example through the formation of Fe(III) or ferric hydroxide in water (Lim and Kim, 2009) and wastewater treatment (Lee *et al.*, 2009). Fe(VI) could partially remove micropollutants by oxidation and subsequent coagulation process to enhance the micropollutants removal from the aqueous phase.

In water treatment, the high oxidative nature of Fe(VI) allows it to possess disinfection ability in addition to the removal of NOM through oxidation-coagulation process (Lim and Kim, 2009). In addition, the formation of non-toxic ferric ion as a by-product of the Fe(VI) reaction favors this chemical to be used in water treatment (Sharma, 2008). The NOM removal performance of Fe(VI) was reported comparable to the traditional coagulants such as alum and ferric sulphate, while a small dose of ferrate as a pretreatment could enhance the removal rate of humic acid by traditional coagulants (Lim and Kim, 2009). At a Fe(VI) dose of 2–46 mg/L, humic and fulvic acids (both with initial concentrations of 10 mg/L) removals were in the range of 21–74% and 48–78%, respectively.

In wastewater treatment, simultaneous removal of micropollutants and phosphate in secondary treated effluent could be achieved with Fe(VI) (Lee *et al.*, 2009). Higher reactivity between Fe(VI) with micropollutants containing phenolic-moiety was reported compared to other micropollutant compounds with amine and olefin moiety. More than 95% removal was achieved for 17β-estradiol, biosphenol A and 17α-ethinylestradiol (containing phenolic-moiety) using 2 mg/L Fe(VI) at pH 7–8 while more than twice this dose was required to achieve similar removal of sulfamethoxazole, diclofenac (containing aniline-moiety) and carbamazepine (containing olefine-moiety). In a study by Lee *et al.* (2009), the depletion of phosphate was only observed after elimination of micropollutants. Hence, a higher Fe(VI) dose would be required to achieve both micropollutants and phosphate removal. It was only at Fe(VI) of 7.5 mg/L that more than 80% phosphate (from 3.5 to less than 0.8 mg PO_4^{3-}-P/L) was achieved (Lee *et al.*, 2009).

The Fe(VI) oxidation reaction with micropollutants followed a second-order rate constant (k) which was shown to increase with a decrease in the reaction pH (Lee *et al.*, 2009). Similarly, under acidic condition, ferrate has three times higher redox potential (which is slightly above ozone). This shows the high potential of Fe(VI) as an oxidant (Lim and Kim, 2009). Under acidic condition, the fraction of Fe(VI) in the protonated form ($HFeO_4^-$) increases (Figure 6.2) (Sharma, 2008). $HFeO_4^-$ species is a strong oxidant as compared with the deprotonated Fe(VI) species ((FeO_4^{2-}), hence contributing to an increase in the oxidation rate at lower pH (Sharma, 2008).

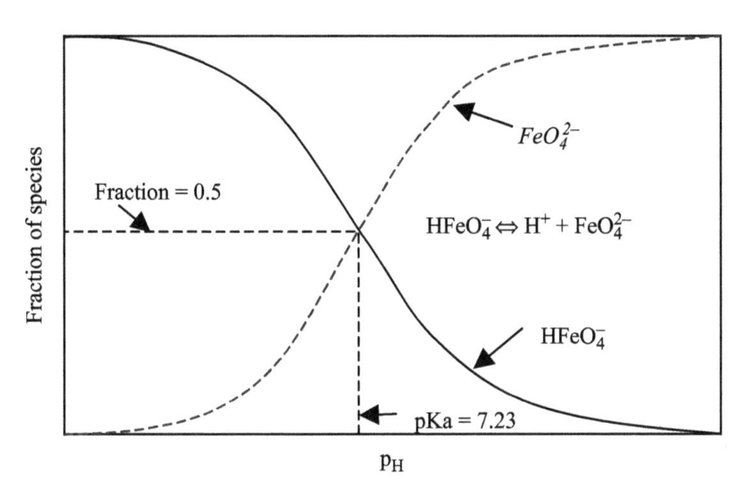

Figure 6.2 Changes in fraction of Fe(VI) species at different pH with pKa = 7.23 at 25°C (extracted from Sharma, 2008)

Fe(VI) also has a higher stability compared to ozone (O_3). Lee *et al.* (2009) showed that Fe(VI) persisted in the secondary treated effluent for more than 30 min compared with O_3 depletion in less than 5 min. The second-order rate constants of Fe(VI) for reaction with micropollutants were demonstrated to be at least 4–5 orders lower than O_3 at pH 7–8 (Lee *et al.*, 2009). A summary on the second order rate constants for selected pollutants with Fe(VI) is given in Table 6.6. A longer reaction time would be required for Fe(VI) as compared with O_3 oxidation. Most pharmaceutical compounds have a half-life lower than 100 s when oxidized with O_3, while the half-life of sulfonamides is more than 2 times longer using Fe(VI) (Sharma, 2008). The application of Fe(VI) in environmental water samples may be dependent on Fe(VI) selectivity in oxidizing the micropollutants. Certain micropollutants with lower oxidation reactions with Fe(VI) such as ibuprofen (first order kinetic constant at 0.9×10^{-1} $M^{-1} \cdot s^{-1}$ at pH 8.0) could be removed by co-precipitation through the formation of ferric oxide (Sharma and Mishra, 2006).

Table 6.6 Second-order rate constant, k for selected micropollutant with Fe(VI) at pH 7

Compound	k ($M^{-1} \cdot s^{-1}$)	Reaction Temperature	Reference
Pharmaceuticals			
Diclofenac	1.3×10^2	23°C	Lee *et al.* (2009)
Carbamezepine	67	23°C	Lee *et al.* (2009)
Sulfamethazine	22.5	25°C	Sharma and Mishra (2006)
Endocrine disruptors			
Bisphenol A	6.4×10^2	23°C	Lee *et al.* (2009)
17β-estradiol	7.6×10^2	23°C	Lee *et al.* (2009)

6.2 MEMBRANE PROCESSES

Membrane processes make use of artificial semi-permeable or porous membranes to separate compounds based either on their size (pressure-driven operation) or on their electrical charge (electrodialysis). Pressure-driven processes can be divided into high-pressure membranes where compound separation occurs mainly by size exclusion and low-pressure membranes where adsorption and other phenomena are responsible for compound separation. Low-pressure membranes have pore sizes in the range of 0.001– 0.1 (ultrafiltration) to 0.1–10 μm (microfiltration). These systems operate under driving pressure below 1,030 KN m^{-2} and are usually utilised to retain large molecules with molecular weight over 5,000 Da such as colloids and oil (Tchobanoglous *et al.*, 2003). Smaller molecules can usually be retained in the process of reverse osmosis (RO) at a driving pressure higher than the osmotic pressure of the solution to be filtered (up to 6,900 KN m^{-2}). This resulted in a higher operating cost for the RO process. On the other hand, electrodialysis works differently by using ion exchange membranes to separate ionic compounds. This is achieved by making an electrical current pass through the solution which causes cations to migrate towards a cation exchange membrane and anions towards an anion exchange membrane. Because cation- and anion-exchange membranes are spaced alternately, this results in alternated compartments of concentrated and diluted solutions.

This section will review the use of microfiltration (MF) and ultrafiltration (UF) as low pressure membrane processes for micropollutant removal, alone or in combination with other physico-chemical techniques (such as particulate activated carbon). It will also cover the incorporation of UF membranes into a biological reactor (known as membrane bioreactor). Rejection of micropollutants by the RO process – a high-pressure membrane – shall be presented but by nanofiltration will not be covered because it is the object of an earlier section of this book. Finally, an insight will be provided into electrodialysis applications for micropollutant removal.

6.2.1 Mechanisms of solute rejection during membrane treatment

Solute rejection in membrane treatment is directly determined by the interactions between the solute and the membrane. These interactions are of different natures, reflecting steric partition, adsorption and charge effects as well as biodegradation phenomena (Figure 6.3). Among these three categories of interactions, steric effects are the most straightforward to understand, directly related to the molecule size (or molecular weight) on the one hand and the membrane pore size on the other hand. If the molecule is bigger than the pores, it will be well rejected. If it is smaller, it will permeate through the membrane. Charge phenomena are related to the fact that most polymeric membranes are negatively charged. As a consequence, repulsive forces will keep away negatively-charged solutes, while positively-charged solutes will be attracted to the membranes. This results in better rejection of negatively-charged molecules by polymeric membranes. The charge of a molecule can be predicted by comparing its p*K*a value to the pH of the solution (Bellona and Drewes, 2005).

On top of the obvious sieving effect of membranes, hydrophobic solutes (i.e., having an octanol-water partitioning coefficient (K_{ow}) higher than 2) can be adsorbed onto the surface of the membrane. This results in increased rejection levels at the beginning of a membrane operation time. However, the adsorption sites on a membrane are not infinite and once the membrane adsorption capacity is reached, hydrophobic solutes will be able to pass through the membrane. This can lead to overestimation of hydrophobic solute rejection if tests are not done under steady state conditions, i.e., when all adsorption sites have been occupied (Kimura *et al.*, 2003a). Furthermore, adsorption of NOM onto the membrane can result in biodegradation phenomena, the importance of which in membrane systems depends on the frequency of backwashing and utilization of sanitisers (Laine *et al.*, 2002). Studies of biofouling have emphasised that the fouling layer acts like a second membrane and the overall impact on micropollutant removal depends on the specific affinity of each layer for the compound studied (Vanoers *et al.,* 1995). If the membrane rejects the solute better than the deposited layer, fouling might result in accumulation of solutes on the membrane surface. Ultimately, this might cause enhanced concentration polarization and decreased solute rejection. However, if the solute is rejected better by the deposited layer than by the membrane, solute rejection might improve.

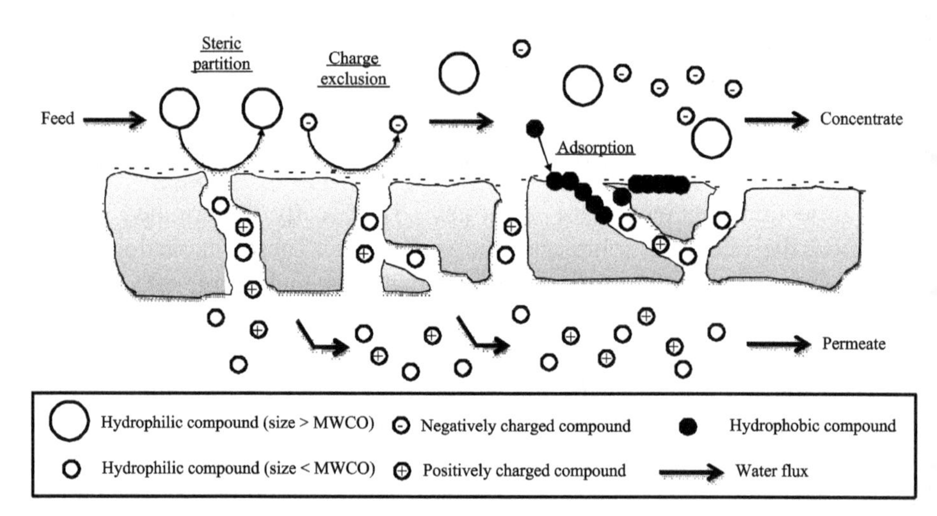

Figure 6.3 Mechanisms of solute partition during treatment by a negatively-charged membrane

6.2.2 Micropollutant removal by microfiltration

MF membranes are a type of low-pressure membranes that are characterized by a pore size in the range of 0.1–10 μm. Even though this pore size is adequate for removal of pathogens in drinking water, it is unlikely that micropollutants (<1 kDa) can be retained by sieving effect. In one study, MF was used as a pre-treatment step before advanced treatment of a water reclamation plant tertiary effluent (Drewes *et al.,* 2003). The fate of trace organic compounds in the form of EDTA, nitrilotriacetic acid and alkylphenolpolyethoxycarboxylates was followed during MF (Osmonics/Desal EW4040F) operation at 3.8 L/min filtrate flow, 7.6 L/min concentrate flow, and a differential pressure of 21.6 kPa. Removals of nitrilotriacetic acid and alkylphenolpolyethoxycarboxylates were found to be ineffective. However, EDTA concentration attained 5.6 μg/L in the permeate as compared to 11.8 μg/L in the tertiary effluent (removal efficiency of 53%). In the absence of an understanding of the mechanism in EDTA removal by the MF membrane, there is a possibility that EDTA was merely adsorbed onto the MF membrane. MF inability to treat pharmaceutically active compounds (PhACs) and hormones was further established in two full-scale wastewater reclamation plants in Australia (Al-Rifai *et al.*, 2007).

6.2.3 Micropollutant removal by ultrafiltration

a) Ultrafiltration alone

Similarly as for MF, the molecular weight cut-off of UF membranes (10–100 kDa) does not allow sieving of most micropollutants. One study considered the role of fouling by NOM on the fate of estradiol and ibuprofen removal by UF membrane (Jermann *et al.*, 2009). Without NOM, retention by an hydrophilic UF membrane (regenerated cellulose, 100 kDa, Ultracel, Millipore, UK) was almost null for both micropollutants, whereas 80% of estradiol and 7% of ibuprofen were retained by a more hydrophobic membrane (polyethersulfone, 100 kDa, Biomax, Millipore, UK). The difference observed between the two types of membranes can be explained by adsorption phenomena on the hydrophobic membrane, as explained above. For ibuprofen, the retention rate was initially higher (25%) but decreased subsequently, which indicates that adsorption was mostly a temporary phenomenon until adsorption sites were saturated. The higher retention of estradiol could also be explained by the more hydrophobic nature of this compound, as compared to ibuprofen, as well as by its negative charge at the pH of the experiment. Most of estradiol could be recovered by sodium hydroxide (pH 12.3), which may have implications during membrane cleaning procedures. The effect of NOM varied depending on the nature of NOM, membrane and micropollutant. NOM could in some cases decrease the retention rate by occupying adsorption sites on the membrane (in the case of humic acids) or in other cases increase the retention of micropollutant by the formed cake layer which acts as a second membrane (in the case of alginate). Another study (Majewska-Nowak *et al.*, 2002) showed that atrazine separation by a polysulfone membrane (Ps, Sartorius) and a composite polysulfone/polyamide membrane (Ds-GS, Osmonics) was maximum at pH 7 (around 20% rejection). In the presence of humic acids at a moderate concentration (up to 20 mg/L), atrazine rejection was improved up to 80% due to atrazine adsorption on humic substances. However, higher doses of humic acids reduced the retention of atrazine, probably due to saturation of the membrane adsorption sites. These two experiments clearly show that the removal of micropollutants by UF membrane is mostly limited to adsorption phenomena, which is not a reliable treatment method on the long term. Surprisingly, one study showed higher rejection of chloroform by UF (93%) than by reverse osmosis or nanofiltration (81 and 75%, respectively) (Waniek *et al.*, 2002). However, the authors were unable to explain the reason of the better performance of UF and it is likely that adsorption is also the main mechanism involved in the retention of chloroform by UF in that experiment.

b) Combination of ultrafiltration and powdered activated carbon

Due to limited removal of micropollutants by UF alone, the combination of UF and powdered activated carbon (PAC) adsorption has attracted the attention of many researchers. The main advantages brought by the combined treatment are increased adsorption kinetics, increased mixing due to UF recirculation loop and flexibility (Laine *et al.*, 2000). However, specific information on the fate of micropollutants during the combined treatment is scarce. One study focused on the fate of three micropollutants (i.e., trichlorethene, tetrachlorethene, atrazine) present in karstic spring water during UF (Aquasource) combined with PAC (Envir-Link MV125) at the pilot scale (Pianta *et al.*, 1998). The rejection level was higher for atrazine (>99% at 5 mg/L of PAC) than for trichlorethene and tetrachlorethene. For the latter two, removal rates of 90% was only made possible by a PAC dose of 22 mg/L. The authors also emphasized on the importance of kinetics and demonstrated that the performance could be improved by reducing the membrane flux, which resulted in increasing contact time between the micropollutants and PAC.

Apart from the dosage of PAC, the way it is administered also has an impact on the efficacy of the UF-PAC treatment systems. In one study, 52% of atrazine was removed by the combined process when PAC was added continuously throughout the filtration cycle, but this percentage increased to 76% when the whole dose of PAC (8 mg/L) was added at the beginning of the cycle (Campos *et al.*, 1998). The authors explained that this is related to the kinetics of atrazine adsorption to PAC, whereby the contact time was being increased when all the PAC was dosed at one time in the beginning. However, another paper reaches the opposite conclusion (Ivancev-Tumbas *et al.*, 2008). In that study, an UF membrane operated in dead-end mode achieved a rejection of p-Nitrophenol by 39% when the PAC dose (10 mg/L) was added at the beginning of the cycle, which was inferior to the performance attained by continuous supply of the same PAC dose (rejection of 75%). The authors attributed the contradictory results to different experimental conditions in terms of membrane type and micropollutant nature and concentration. It is also possible that a single dosing of PAC resulted in the formation of a carbon monolayer inside the capillaries of the UF membrane that slowed the adsorption kinetics of the micropollutant onto the PAC and also reduced the adsorption capacity of the membrane.

c) Combination of ultrafiltration and biological module (membrane bioreactor)

An alternative to PAC in order to improve the performance of UF is the addition of a biological compartment to the UF membrane, resulting in a process known as membrane bioreactor (MBR). In an MBR, the biological component is responsible for the biological degradation of waste compounds while the membrane module accomplishes the solids-liquids separation (Choi and Ng, 2008; Ng and Hermanowicz, 2005). MBRs can be either submerged, i.e., membrane modules are directly immersed in the activated sludge tank, or side-stream, in which case membrane filtration is achieved outside of the activated sludge tank. MBR has attracted increasing interest in recent years in wastewater treatment because of the emergence of low-cost membranes with lower pressure requirements and higher permeate fluxes. One specific advantage that makes MBR attractive for micropollutant removal is the long sludge retention time (SRT) allowed by the technology (Tan *et al.*, 2008).

An early study focused on the removal of polar sulphur aromatic micropollutants: naphthalene sulfonates and benzothiazoles in a large-scale side-stream MBR treating tannery wastewater (Reemtsma *et al.*, 2002). Naphtalene monosulfonates were removed to a large extent (>99%) but disulfonate removal averaged only 44%. The same profile and variability depending on every specific compound were observed for the family of benzothiazoles with 2-mercaptobenzothiazole being almost entirely removed but 2-aminobenzothiazole being almost unaltered. It was found that a fraction of 2-mercaptobenzothiazole was oxidized into a byproduct benzothiazole-2-sulfonic acid that did not seem to be further altered. In that sense, MBR technology did not seem to produce results largely different from the conventional activated sludge process for some micropollutants. The same conclusion was reached in a comprehensive study comparing the fate of a variety of PPCPs and endocrine disrupting compounds (EDCs) in an MBR pilot plant and three conventional activated sludge wastewater treatment plants (Clara *et al.*, 2005). Most of the compounds analysed were removed to the extent of 50 to 60% with the notable exception of bisphenol-A, ibuprofen and bezafibrate, for which the removal rate exceeded 90%. In all cases, no significant difference was observed between the MBR and conventional activated sludge process, showing that the UF membrane does not allow further removal of some micropollutants.

Excellent MBR removal of ibuprofen (up to 98%) was confirmed in another study along with naproxen (84%) and erythromycin (91%) and these performances were attributed by the authors to the exceptionally long SRT (44–72 d) achieved in their pilot-scale immersed MBR (Reif *et al.*, 2007).

In addition, sulfamethoxazole and musk fragrances (galaxolide, tonalide, celestolide) were moderately removed ($>50\%$) probably due to partial adsorption on the biomass. On the other hand, carbamazepine, diazepam, diclofenac and trimethoprim were poorly removed ($<10\%$) due to poorer biodegradability. Another study confirmed the possibility of achieving good treatment of bisphenol-A (90%) due to both biodegradation and adsorption phenomena (Nghiem *et al.,* 2007). On the contrary, sulfamethoxazole removal was solely attributed to biodegradation, which can explain the poorer rejection level of this hydrophilic compound (50%). In the case of a decentralized MBR treating domestic wastewater spiked with PhACs, sulfonamide antibiotics, ibuprofen, bezafibrate, estrone and 17α-ethinylestradiol appeared to be easily biodegradable mostly under aerobic or anoxic conditions in the presence of nitrate (Abegglen *et al.,* 2009). However, macrolide antibiotics appeared poorly biodegradable and their elimination was mostly due to adsorption phenomenon. Another study confirmed the good biodegradability ($>99\%$) of 17α-ethinylestradiol in an MBR enriched with nitrifiers (De Gusseme *et al.,* 2009).

The fate of metals in MBRs has also gained interest in the recent years. One paper studied the impact of SRT on the removal of various metals in an immersed MBR (Innocenti *et al.,* 2002). Again, high variability was observed with Ag, Cd and Sn being largely removed ($>99\%$); Cu, Hg and Pb moderately removed ($>50\%$); but B, Se and As poorly treated ($<50\%$). Moreover, for As, Pb, Se and B, increasing SRT did not improve the removal of these species, which seemed to indicated that adsorption on the biomass was not involved in these metals removal. It is more likely that their charge effects with the UF membrane were directly linked to their rejection. Another study showed slightly different results with higher removal of As (65%), Hg and Cu (both over 90%) but on the other hand, slightly lower performance on Cd ($>50\%$ removal) (Fatone *et al.,* 2005). In the same study, the fate of a variety of polycyclic aromatic hydrocarbon (PAHs) was also studied, for which moderate removal in the range of 58 to 76% was observed.

Due to degradation in MBR being largely dominated by the biological component, bioaugmentation was shown to help achieving higher removal of compounds such as nonylphenol (Cirja *et al.,* 2009). In some cases, MBR can be further combined with PAC for improved micropollutant removal, such as pesticides (Laine *et al.,* 2000). A summary of these studies and additional data on micropollutant rejection by MBR technology are given in Table 6.7.

Table 6.7 Removal of micropollutants by the MBR technology

Classification	Micropollutant	Removal Efficiency (%)	Reference
Benzothiazole	2-Hydroxybenzothiazole	0	(Reemtsma *et al.*, 2002)
	2-Mercaptobenzothiazole	>99	(Reemtsma *et al.*, 2002)
	Benzothiazole	5	(Reemtsma *et al.*, 2002)
	Benzothiazole-2-sulfonic acid	0	(Reemtsma *et al.*, 2002)
	Methylthiobenzothiazole	50	(Reemtsma *et al.*, 2002)
EDC	Nonylphenol	91	(Clara *et al.*, 2005)
	Nonylphenol diethoxylate	94	(Clara *et al.*, 2005)
	Nonylphenol monoethoxylate	99	(Clara *et al.*, 2005)
	Nonylphenoxyacetic acid	0	(Clara *et al.*, 2005)
	Nonylphenoxyethoxyacetic acid	0	(Clara *et al.*, 2005)
	Octylphenol	>99	(Clara *et al.*, 2005)
	Octylphenol diethoxylate	>99	(Clara *et al.*, 2005)
	Octylphenol monoethoxylate	>99	(Clara *et al.*, 2005)
	17α-Ethinylestradiol	99	(De Gusseme *et al.*, 2009)
	Bisphenol-A	99	(Clara *et al.*, 2005)
		90	(Nghiem *et al.*, 2007)
Metal	Ag	>99	(Innocenti *et al.*, 2002)
	Al	>99	(Innocenti *et al.*, 2002)
		>96	(Fatone *et al.*, 2005)
	As	65	(Fatone *et al.*, 2005)
		35	(Innocenti *et al.*, 2002)

(*continued*)

Table 6.7 (*continued*)

Classification	Micropollutant	Removal Efficiency (%)	Reference
	B	30	(Innocenti *et al.*, 2002)
	Ba	85	(Innocenti *et al.*, 2002)
	Cd	>99	(Innocenti *et al.*, 2002)
		>50	(Fatone *et al.*, 2005)
	Co	80	(Innocenti *et al.*, 2002)
	Cr	75	(Fatone *et al.*, 2005)
Metal	Cu	96	(Fatone *et al.*, 2005)
		85	(Innocenti *et al.*, 2002)
	Fe	>97	(Fatone *et al.*, 2005)
		95	(Innocenti *et al.*, 2002)
	Hg	99	(Innocenti *et al.*, 2002)
		94	(Fatone *et al.*, 2005)
	Mn	80	(Innocenti *et al.*, 2002)
	Ni	79	(Fatone *et al.*, 2005)
		60	(Innocenti *et al.*, 2002)
	Pb	74	(Fatone *et al.*, 2005)
		60	(Innocenti *et al.*, 2002)
	Se	30	(Innocenti *et al.*, 2002)
	V	90	(Innocenti *et al.*, 2002)
	Zn	>90	(Fatone *et al.*, 2005)
		80	(Innocenti *et al.*, 2002)

(*continued*)

Table 6.7 (*continued*)

Classification	Micropollutant	Removal Efficiency (%)	Reference
Naphthalene sulfonate	1,7-Naphthalene disulfonate	40	(Reemtsma *et al.*, 2002)
	2,6-Naphthalene disulfonate	90	(Reemtsma *et al.*, 2002)
	2,7-Naphthalene disulfonate	0	(Reemtsma *et al.*, 2002)
	1-Naphthalene monosulfonate	>99	(Reemtsma *et al.*, 2002)
	2-Naphthalene monosulfonate	>99	(Reemtsma *et al.*, 2002)
	1,5-Naphthalene disulfonate	0	(Reemtsma *et al.*, 2002)
	1,6-Naphthalene disulfonate	80	(Reemtsma *et al.*, 2002)
PAH	Acenafthene	76	(Fatone *et al.*, 2005)
	Acenafthylene	>61	(Fatone *et al.*, 2005)
	Fenanthrene	71	(Fatone *et al.*, 2005)
	Fluoranthene	65	(Fatone *et al.*, 2005)
	Fluorene	>54	(Fatone *et al.*, 2005)
	Nafthalene	66	(Fatone *et al.*, 2005)
	Pyrene	58	(Fatone *et al.*, 2005)
PhAC	Diclofenac	50	(Clara *et al.*, 2005)
		0	(Reif *et al.*, 2007)
	Ibuprofen	99	(Reif *et al.*, 2007)
		99	(Clara *et al.*, 2005)
	Sulfamethoazole	61	(Clara *et al.*, 2005)
		55	(Reif *et al.*, 2007)
		50	(Nghiem *et al.*, 2007)
	Carbamazepine	12	(Clara *et al.*, 2005)
		10	(Reif *et al.*, 2007)

(*continued*)

Table 6.7 (*continued*)

Classification	Micropollutant	Removal Efficiency (%)	Reference
PPCP	Bezafibrate	96	(Clara *et al.*, 2005)
	Diazepam	25	(Reif *et al.*, 2007)
	Erythromycin	90	(Reif *et al.*, 2007)
	Galaxolide	60	(Reif *et al.*, 2007)
		92	(Clara *et al.*, 2005)
	Naproxen	85	(Reif *et al.*, 2007)
	Roxithromycin	>99	(Clara *et al.*, 2005)
		80	(Reif *et al.*, 2007)
	Tonalide	85	(Clara *et al.*, 2005)
		40	(Reif *et al.*, 2007)
	Trimethoprim	30	(Reif *et al.*, 2007)
	Celestolide	50	(Reif *et al.*, 2007)

6.2.4 Micropollutant removal by reverse osmosis

RO membranes are characterized by having "pores" with size ranging between 0.22 and 0.44 nm (Kosutic and Kunst, 2002). As a consequence, RO – like nanofiltration – has the potential to result in partial or highly effective removal of most micropollutants (Jermann *et al.*, 2009). However, this is achieved at the cost of higher energy consumption (Jones *et al.*, 2007) and the development of ultra-low pressure reverse osmosis membranes, in which the water flux is facilitated by a hydrophilic support layer is of major concern (Ozaki and Li, 2002). In addition to steric effects, solute transport in RO membranes is influenced by diffusion and electrostatic phenomena throughout membrane pores and, as a consequence, the tightest pores do not always result in the best rejection of micropollutants (Kosutic and Kunst, 2002). In their study, this was notably the case for 2-butanone, better removed by looser membranes, and for the pesticide triadimefon that showed reduced rejection pattern (58–82%) as compared to atrazine (80–99%) in spite of its higher molecular width and weight. The main difference between RO and nanofiltration membranes is in terms of selectivity – RO being designed to remove all ions (including monovalent ions) wheareas nanofiltration are selective towards multivalent ions (Li, 2007).

Early studies of micropollutant removal by RO date back to the 1980s. It was found at that time that most metals were largely rejected by RO (>75%). However, organic micropollutants in the form of trihalomethanes, dichloromethane and alkylphenols were poorly removed (Hrubec *et al.,* 1983). Later on, a study investigated the performance of different types of RO membranes for micropollutant removal including several pesticides and chlorophenols (Hofman *et al.,* 1997). Ultra-low pressure RO membranes were found to compete well with polyamide membranes, almost all compounds tested being rejected below their detection level. However, cellulose-acetate displayed poorer performance.

In a very comprehensive study, the fate of DBPs, EDCs and PhACs by RO filtration was investigated (Kimura *et al.,* 2003b). In their study, adsorption was found to be an early mechanism in micropollutant removal but only temporarily after what steric and charge effects were mostly responsible for micropollutant removal. Notably, negatively charged molecules were found to be better rejected (>90%) than non-charged compounds (<90%). The exception was for the EDC Bisphenol-A that, in spite of being neutral (uncharged), could be removed up to 99% using an ultra-low pressure RO membrane (RO-XLE, Film-Tec, Vista, CA). In a later study, the overall better performance of this polyamide membrane as compared to a cellulose acetate one (SC-3100, Toray) was confirmed with the notable exception of the pharmaceutical compound sulfamethoxazole (Kimura *et al.,* 2004). The good performance of this polyamide membrane was also demonstrated on a variety of antibiotics that showed rejection rates of over 97% (Kosutic *et al.,* 2007).

The possibility of treating efficiently EDCs and PhACs over 90% by RO was further showcased in another study, in which most of the compounds studied were rejected to below detection levels (<25 ng/L) (Comerton *et al.,* 2008). Another study of two full-scale wastewater treatment plants further showed that RO was the most effective step for the treatment of PhACs and non-steroidal estrogenic compounds (Al-Rifai *et al.,* 2007). Total rejection of the antibiotic triclosan was also showcased during a 10-h filtration testing through a BW-30 (Dow FilmTec, Minneapolis, MN) RO membrane (Nghiem and Coleman, 2008). However, caffeine was found in the permeate of two full-scale RO facilities (Drewes *et al.,* 2005). This could be explained by its hydrophilic and non-ionic nature limiting its rejection to size exclusion effects. Similarly, the same study showed limited rejection of chloroform in the range of 50 to 85%.

As for other types of membranes, fouling by NOM was shown to increase the retention of hydrophobic non-ionic molecules by acting as a second membrane. However, in some cases, fouling by NOM resulted in membrane swelling and decreased rejection capacities (Xu *et al.,* 2006). Again, the nature of the membrane was found to be of importance with cellulose triacetate and ultra-low pressure RO membranes being affected by fouling more than thin film composite membranes. The nature of the foulant is also important, with colloidal fouling being known to be particularly responsible for a decrease in water flux and permeate quality (Ng and Elimelech, 2004). In that study, colloidal fouling caused a decrease in RO (LFC-1, Hydranautics, Oceanside, CA) rejection of hormones (estradiol and progesterone) and, contrary as what was observed for salts and inert organic solutes, hormone rejection continued to decrease even after the water flux had stabilized. As a consequence, hormone removal constantly decreased from over 95% initially to values in the range of 75 to 85% in the presence of colloids (200 mg/L) after 110 h. This shows that hormone rejection by RO membrane was mostly due to adsorption phenomena, whereas larger organic molecules were removed by steric effects. In addition, in that crossflow experiment, hormone rejection was found to be about 10% higher when the height of the membrane filtration cell channel was reduced by a factor of 2, showing the importance of shear forces. Improved hormone rejection at higher shear rates could be explained by decreased concentration polarization. Comprehensive data on micropollutant removal by RO is provided in Table 6.8.

6.2.5 Electrodialysis

Recently, urine treatment by electrodialysis has brought to the light the potential of such membranes for micropollutant removal. Even though ethinylestradiol, a major compound excreted in urine, was found to be fully rejected in the long-term operation of laboratory electrodialysis membranes (Mega a.s., Prague, Czech Republic), good rejection was only temporary for other compounds such as diclofenac, carbamazepine, propranolol and ibuprofen, showing that most of the rejection mechanism was caused by adsorption for these molecules (Pronk *et al.,* 2006). Overall, electrodialysis showed the best performance out of five treatment processes for urine, the other four being sequencing batch reactor, nanofiltration, struvite precipitation and ozone treatment, achieving up to 99.7% removal efficiency of estrogenicity (Escher *et al.,* 2006).

Table 6.8 Removal of micropollutants by RO

Classification	Name of micropollutant	Removal (%)	Reference
Acid	Acetic acid	>99	(Ozaki and Li, 2002)
	Dichloroacetic acid	95	(Kimura *et al.*, 2003b)
Alcohol	2-Propanol	86	(Kosutic *et al.*, 2007)
	Benzyl alcohol	85	(Ozaki and Li, 2002)
	Erythritol	93	(Ng and Elimelech, 2004)
	Ethyl alcohol	40	(Ozaki and Li, 2002)
	Ethylene glycol	43	(Ng and Elimelech, 2004)
		50	(Ozaki and Li, 2002)
	Glycerol	92	(Kosutic *et al.*, 2007)
		93	(Ng and Elimelech, 2004)
	Methyl alcohol	25	(Ozaki and Li, 2002)
	o-Nitrophenol	90	(Ozaki and Li, 2002)
	Phenol	75	(Ozaki and Li, 2002)
	p-Nitrophenol	95	(Ozaki and Li, 2002)
	Triethylene glycol	90	(Ozaki and Li, 2002)
Alkanes	Alkanes	90	(Hrubec *et al.*, 1983)
Alkylbenzenes	C_1 Alkylbenzenes	8	(Hrubec *et al.*, 1983)
	C_2 Alkylbenzenes	8	(Hrubec *et al.*, 1983)
	C_3 Alkylbenzenes	80	(Hrubec *et al.*, 1983)
	C_4 Alkylbenzenes	85	(Hrubec *et al.*, 1983)
	C_5 Alkylbenzenes	90	(Hrubec *et al.*, 1983)
Alkylindanes	Alkylindanes	90	(Hrubec *et al.*, 1983)
Alkylnaphthalenes	Alkylnaphthalenes	15	(Hrubec *et al.*, 1983)
Alkylphenols	Alkylphenols	70	(Hrubec *et al.*, 1983)
Antibiotic	Triclosan	>99	(Nghiem and Coleman, 2008)
Antiepileptic	Primidone	>99	(Drewes *et al.*, 2005)
Aromatic acid	2,4-Dihydroxybenzoic	90	(Xu *et al.*, 2006)
Aromatic acid	2-Naphthalenesulfonic acid	90	(Xu *et al.*, 2006)

(*continued*)

Table 6.8 (*continued*)

Classification	Name of micropollutant	Removal (%)	Reference
Aromatic amine	Aniline	75	(Ozaki and Li, 2002)
Aromatic hydrocarbon	Benzene	8	(Hrubec *et al.*, 1983)
Carbohydrate	Glucose	95	(Ozaki and Li, 2002)
	Xylose	95	(Ng and Elimelech, 2004)
Chlorinated aliphatic	Dichloromethane	0	(Hrubec *et al.*, 1983)
Chlorophenol	2,3,6-Trichlorophenol	>99	(Hofman *et al.*, 1997)
	2,3-Dichlorophenol	95	(Ozaki and Li, 2002)
	2,4,5-Trichlorophenol	95	(Ozaki and Li, 2002)
	2,4,6-Trichlorophenol	>99	(Hofman *et al.*, 1997)
	2,4-Dichlorophenol	90	(Ozaki and Li, 2002)
		>99	(Hofman *et al.*, 1997)
	2,4-Dinitrophenol	95	(Ozaki and Li, 2002)
	2,6-Dichlorophenol	>99	(Hofman *et al.*, 1997)
	4-Chlorophenol	65	(Ozaki and Li, 2002)
	Pentachlorophenol	99	(Ozaki and Li, 2002)
		>99	(Hofman *et al.*, 1997)
Chlorophosphate	Chlorophosphates	60	(Hrubec *et al.*, 1983)
Cyclic hydrocarbon	Cyclic hydrocarbons	80	(Hrubec *et al.*, 1983)
	Cyclohexanone	98	(Kosutic *et al.*, 2007)
Disinfection by-product	Bromoform	95	(Drewes *et al.*, 2005)
	Trichloroacetic acid	96	(Kimura *et al.*, 2003b)
EDC	Progesterone	85	(Ng and Elimelech, 2004)
	Testosterone	>91	(Drewes *et al.*, 2005)
	Diethylstilbesterol	>99	(Comerton *et al.*, 2008)
	Equilin	98	(Comerton *et al.*, 2008)
	DEET	92	(Comerton *et al.*, 2008)
EDC	Estrone	98	(Comerton *et al.*, 2008)

(*continued*)

Table 6.8 (*continued*)

Classification	Name of micropollutant	Removal (%)	Reference
	17α-Estradiol	97	(Comerton *et al.*, 2008)
	17β-Estradiol	83	(Kimura *et al.*, 2004)
		96	(Comerton *et al.*, 2008)
		>93	(Drewes *et al.*, 2005)
	Estradiol	90	(Ng and Elimelech, 2004)
	Estriol	91	(Comerton *et al.*, 2008)
		>80	(Drewes *et al.*, 2005)
	17α-Ethynyl estradiol	98	(Comerton *et al.*, 2008)
	Alachlor (Lasso)	95	(Comerton *et al.*, 2008)
	Atraton	92	(Comerton *et al.*, 2008)
	Carbaryl	79	(Kimura *et al.*, 2004)
	Metolachlor	95	(Comerton *et al.*, 2008)
	Bisphenol-A	66	(Drewes *et al.*, 2005)
		83	(Kimura *et al.*, 2004)
		95	(Comerton *et al.*, 2008)
		99	(Kimura *et al.*, 2003b)
	Oxybenzone	>99	(Comerton *et al.*, 2008)
Flame retardant	Tris(1,3-dichloro-2-propyl)-phosphate	>99	(Drewes *et al.*, 2005)
	Tris(2-chloroethyl)-phosphate	>99	(Drewes *et al.*, 2005)
	Tris(2-chloroisopropyl)-phosphate	>99	(Drewes *et al.*, 2005)
Metal	As	88	(Hrubec *et al.*, 1983)
	Cd	75	(Hrubec *et al.*, 1983)
	Cr	72	(Hrubec *et al.*, 1983)
	Cu	72	(Hrubec *et al.*, 1983)
	Hg	0	(Hrubec *et al.*, 1983)
Metal	Mo	71	(Hrubec *et al.*, 1983)
	Ni	85	(Hrubec *et al.*, 1983)
	Pb	85	(Hrubec *et al.*, 1983)

(*continued*)

Table 6.8 (*continued*)

Classification	Name of micropollutant	Removal (%)	Reference
	Zn	75	(Hrubec *et al.*, 1983)
PAH	Hydronaphthalenes	75	(Hrubec *et al.*, 1983)
	Naphthalene	15	(Hrubec *et al.*, 1983)
Pesticide	Atrazine	99	(Kosutic and Kunst, 2002)
		>99	(Hofman *et al.*, 1997)
	Bentazon	>99	(Hofman *et al.*, 1997)
	Diuron	>99	(Hofman *et al.*, 1997)
	DNOC	>99	(Hofman *et al.*, 1997)
	MCPA	94	(Kosutic and Kunst, 2002)
		95	(Ozaki and Li, 2002)
		>99	(Hofman *et al.*, 1997)
	Mecoprop	>99	(Hofman *et al.*, 1997)
	Metalaxyl	>99	(Hofman *et al.*, 1997)
	Metamitron	>99	(Hofman *et al.*, 1997)
	Metribuzin	>99	(Hofman *et al.*, 1997)
	Pirimicarb	>99	(Hofman *et al.*, 1997)
	Propham	97	(Kosutic and Kunst, 2002)
	Simazin	>99	(Hofman *et al.*, 1997)
	Triadimefon	83	(Kosutic and Kunst, 2002)
	Vinchlozolin	>99	(Hofman *et al.*, 1997)
Petrochemical	1,2-ethanediol	62	(Kosutic and Kunst, 2002)
	2-Butanone	66	(Kosutic *et al.*, 2007)
		78	(Kosutic and Kunst, 2002)
	Ethylacetate	75	(Kosutic and Kunst, 2002)
	Formaldehyde	31	(Kosutic and Kunst, 2002)

(*continued*)

Table 6.8 (*continued*)

Classification	Name of micropollutant	Removal (%)	Reference
	Indane	90	(Hrubec *et al.*, 1983)
PhAC	Diclofenac	95	(Kimura *et al.*, 2003b)
	Isopropylantipyrine	78	(Kimura *et al.*, 2004)
	Phenacetine	71	(Kimura *et al.*, 2003b)
		74	(Kimura *et al.*, 2004)
	Primidone	84	(Kimura *et al.*, 2003b)
		87	(Kimura *et al.*, 2004)
		90	(Xu *et al.*, 2006)
	Carbadox	90	(Comerton *et al.*, 2008)
	Enrofloxacin	99	(Kosutic *et al.*, 2007)
	Levamisole	99	(Kosutic *et al.*, 2007)
	MBIK	97	(Kosutic *et al.*, 2007)
	Oxytetracycline	99	(Kosutic *et al.*, 2007)
	Praziquantel	99	(Kosutic *et al.*, 2007)
	Sulfachloropyridazine	94	(Comerton *et al.*, 2008)
	Sulfadiazine	99	(Kosutic *et al.*, 2007)
	Sulfaguanidine	99	(Kosutic *et al.*, 2007)
	Sulfamerazine	88	(Comerton *et al.*, 2008)
	Sulfamethazine	99	(Kosutic *et al.*, 2007)
	Sulfamethizole	93	(Comerton *et al.*, 2008)
	Sulfamethoxazole	70	(Kimura *et al.*, 2004)
		94	(Comerton *et al.*, 2008)
	Trimethoprim	99	(Kosutic *et al.*, 2007)
	Carbamazepine	91	(Kimura *et al.*, 2004)
		91	(Comerton *et al.*, 2008)
	Acetaminophen	82	(Comerton *et al.*, 2008)
	Gemfibrozil	98	(Comerton *et al.*, 2008)
	Caffeine	70	(Kimura *et al.*, 2004)
		87	(Comerton *et al.*, 2008)
Phenylphenol	4-Phenylphenol	61	(Kimura *et al.*, 2004)
Phosphate	Organic phosphates	60	(Hrubec *et al.*, 1983)

(*continued*)

Table 6.8 (*continued*)

Classification	Name of micropollutant	Removal (%)	Reference
Phthalate	Phthalates	55	(Hrubec *et al.*, 1983)
Sulphonamide	Sulphonamides	10	(Hrubec *et al.*, 1983)
Surrogate	2-Naphthol	43	(Kimura *et al.*, 2003b)
	2-Naphthol	57	(Kimura *et al.*, 2004)
	9-Anthracene carbonic acid	96	(Kimura *et al.*, 2003b)
	Salicylic acid	92	(Kimura *et al.*, 2003b)
Metabolite	Urea	30	(Ozaki and Li, 2002)
Trihalomethane	Bromoform	80	(Xu *et al.*, 2006)
		13	(Hrubec *et al.*, 1983)
	Bromodichloromethane	6	(Hrubec *et al.*, 1983)
	Dibromochloroethane	>99	(Hrubec *et al.*, 1983)
	Trichloroethylene	80	(Xu *et al.*, 2006)
	Chloroform	5	(Hrubec *et al.*, 1983)
		25	(Xu *et al.*, 2006)
		85	(Drewes *et al.*, 2005)

6.3 REFERENCES

Abegglen, C., Joss, A., McArdell, C. S., Fink, G., Schlusener, M. P., Ternes, T. A. and Siegrist, H. (2009) The fate of selected micropollutants in a single-house MBR. *Water Res.* **43**, 2036–2046.

Al-Rifai, J. H., Gabelish, C. L. and Schafer, A. I. (2007) Occurrence of pharmaceutically active and non-steroidal estrogenic compounds in three different wastewater recycling schemes in Australia. *Chemosphere* **69**, 803–815.

Asakura, H. and Matsuto, T. (2009) Experimental study of behaviour of endocrine-disrupting chemicals in leachate treatment process and evaluation of removal efficiency. *Waste Manage.* **29**, 1852–1859.

Bellona, C. and Drewes, J. E. (2005) The role of membrane surface charge and solute physico-chemical properties in the rejection of organic acids by NF membranes. *J. Membrane Sci.* **249**, 227–234.

Bodzek, M. and Dudziak, M. (2006) Elimination of steroidal sex hormones by conventional water treatment and membrane processes. *Desalination* **198**, 24–32.

Campos, C., Marinas, B. J., Snoeyink, V. L., Baudin, I. and Laine, J. M. (1998) Adsorption of trace organic compounds in CRISTAL (R) processes Conference on Membranes in Drinking and Industrial Water Production, Amsterdam, Netherlands, pp. 265–271.

Carballa, M., Omil, F. and Lema, J. M. (2005) Removal of cosmetic ingredients and pharmaceuticals in sewage primary treatment. *Water Res.* **39**, 4790–4796.

Choi, J. H. and Ng, H. Y. (2008) Effect of membrane type and material on performance of a submerged membrane bioreactor. *Chemosphere*, **71**, 853–859.

Cirja, M., Hommes, G., Ivashechkin, P., Prell, J., Schaffer, A., Corvini, P. F. X. and Lenz, M. (2009) Impact of bio-augmentation with Sphingomonas sp strain TTNP3 in membrane bioreactors degrading nonylphenol. *Appl. Microbiol. Biotechnol.* **84**, 183–189.

Clara, M., Strenn, B., Gans, O., Martinez, E., Kreuzinger, N. and Kroiss, H. (2005) Removal of selected pharmaceuticals, fragrances and endocrine disrupting compounds in a membrane bioreactor and conventional wastewater treatment plants. *Water Res.* **39**, 4797–4807.

Comerton, A. M., Andrews, R. C., Bagley, D. M. and Hao, C. Y. (2008) The rejection of endocrine disrupting and pharmaceutically active compounds by NF and RO membranes as a function of compound a water matrix properties. *J. Memb. Sci.* **313**, 323–335.

Deborde, M., Rabouan, S., Duguet, J. P. and Legube, B. (2005) Kinetics of aqueous ozone-induced oxidation of some endocrine disruptors. *Environ. Sci. Technol.* **39**, 6068–6092.

De Gusseme, B., Pycke, B., Hennebel, T., Marcoen, A., Vlaeminck, S. E., Noppe, H., Boon, N. and Verstraete, W. (2009) Biological removal of 17 alpha-ethinylestradiol by a nitrifier enrichment culture in a membrane bioreactor. *Water Res.* **43**, 2493–2503.

Drewes, J. E., Bellona, C., Oedekoven, M., Xu, P., Kim, T. U. and Amy, G. (2005) Rejection of wastewater-derived micropollutants in high-pressure membrane applications leading to indirect potable reuse. *Environ. Prog.* **24**, 400–409.

Drewes, J. E., Reinhard, M. and Fox, P. (2003) Comparing microfiltration-reverse osmosis and soil-aquifer treatment for indirect potable reuse of water. *Water Res.* **37**, 3612–3621.

Edwards, M., Scardina, P. and McNeil, L. S. (2003) Enhanced coagulation impacts on water treatment plant infrastructure. Awwa Research Foundation. IWA Publishing, London, UK.

Escher, B. I., Pronk, W., Suter, M. J .F. and Maurer, M. (2006) Monitoring the removal efficiency of pharmaceuticals and hormones in different treatment processes of source-separated urine with bioassays. *Environ. Sci. Technol.* **40**, 5095–5101.

Fatone, F., Bolzonella, D., Battistoni, P. and Cecchi, F. (2005) Removal of nutrients and micropollutants treating low loaded wastewaters in a membrane bioreactor operating the automatic alternate-cycles process European Conference on Desalination and the Environment, St Margherita, Italy, pp. 395–405.

Freese, S. D., Nozaic, D. J., Pryor, M. J., Rajogopaul, R., Trollip, D. L. and Smith, R. A. (2001) Enhanced coagulation: a viable option to advance treatment technologies in the South African context. *Water Sci. Technol: Water Supply* **1**(1), 33–41.

Hemond, H. F. and Fechner-Levy, E. J. (2000) Chemical Fate and Trasnport in the Environment. 2nd Edition. Academic Press, San Diego, USA.

Hofman, J., Beerendonk, E. F., Folmer, H. C. and Kruithof, J. C. (1997) Removal of pesticides and other micropollutants with cellulose-acetate, polyamide and ultra-low pressure reverse osmosis membranes Workshop on Membranes in Drinking Water Production – Technical Innovations and Health Aspects, Laquila, Italy, pp. 209–214.

Hrubec, J., Vankreijl, C. F., Morra, C. F. H. and Slooff, W. (1983) Treatment of municipal wastewater by reverse-osmosis and activated-carbon-removal of organic micro-pollutants and reduction of toxicity. *Sci. Tot. Environ.* **27**, 71–88.

Innocenti, L., Bolzonella, D., Pavan, P. and Cecchi, F. (2002) Effect of sludge age on the performance of a membrane bioreactor: influence on nutrient and metals removal International Congress on Membranes and Membrane Processes (ICOM), Taulouse, France, pp. 467–474.

Ivancev-Tumbas, I., Hobby, R., Kuchle, B., Panglisch, S. and Gimbel, R. (2008) p-Nitrophenol removal by combination of powdered activated carbon adsorption and ultrafiltration – comparison of different operational modes. *Water Res.* **42**, 4117–4124.

Jermann, D., Pronk, W., Boller, M. and Schafer, A. I. (2009) The role of NOM fouling for the retention of estradiol and ibuprofen during ultrafiltration. *J. Memb. Sci.* **329**, 75–84.

Jones, O. A. H., Green, P. G., Voulvoulis, N. and Lester, J. N. (2007) Questioning the excessive use of advanced treatment to remove organic micropollutants from wastewater. *Environ. Sci. Technol.* **41**, 5085–5089.

Kimura, K., Amy, G., Drewes, J. and Watanabe, Y. (2003a) Adsorption of hydrophobic compounds onto NF/RO membranes: An artifact leading to overestimation of rejection. *J. Memb. Sci.* **221**, 89–101.

Kimura, K., Amy, G., Drewes, J. E., Heberer, T., Kim, T. U. and Watanabe, Y. (2003b) Rejection of organic micropollutants (disinfection by-products, endocrine disrupting compounds, and pharmaceutically active compounds) by NF/RO membranes. *J. Memb. Sci.* **227**, 113–121.

Kimura, K., Toshima, S., Amy, G. and Watanabe, Y. (2004) Rejection of neutral endocrine disrupting compounds (EDCs) and pharmaceutical active compounds (PhACs) by RO membranes. *J. Memb. Sci.* **245**, 71–78.

Kosutic, K., Dolar, D., Asperger, D. and Kunst, B. (2007) Removal of antibiotics from a model wastewater by RO/NF membranes. *Sep. Purif. Technol.* **53**, 244–249.

Kosutic, K. and Kunst, B. (2002) Removal of organics from aqueous solutions by commercial RO and NF membranes of characterized porosities. *Desalination* **142**, 47–56.

Lai, K. M., Johnson, K. L., Scrimshaw, M. D. and Lester, J. N. (2000) Binding of waterborne steroid estrogens to solid phases in river and estuarine systems. *Environ. Sci. Technol.* **34**, 3890–3894.

Laine, J. M., Campos, C., Baudin, I. and Janex, M. L. (2002) Understanding membrane fouling: A review of over a decade of research. in: G. Hagmeyer, J.C. Schipper, R. Gimbel (Eds.), 3rd International Conference on Membranes in Drinking and Industrial Water Production, Mulheim, Germany, pp. 155–164.

Laine, J. M., Vial, D. and Moulart, P. (2000) Status after 10 years of operation – overview of UF technology today Conference on Membranes in Drinking and Industrial Water Production, Paris, France, pp. 17–25.

Lee, Y., Zimmermann, S. G., Kieu, A. T. and Von Gunten, U. (2009) Ferrate (Fe(VI)) application for municipal wastewater treatment: A novel process for simultaneous micropollutant oxidation and phosphate removal. *Environ. Sci. Technol.* **43**, 3831–3838.

Li, K. (2007) Ceramic membranes for separation and reaction. John Wiley and Sons.

Lim, M. and Kim, M. J. (2009) Removal of natural organic matter from river water using potassium ferrate (VI). *Water, Air, and Soil Poll.* **200**, 181–189.

Majewska-Nowak, K., Kabsch-Korbutowicz, M. and Dodz, M. (2002) Effects of natural organic matter on atrazine rejection by pressure driven membrane processes International Congress on Membranes and Membrane Processes (ICOM), Taulouse, France, pp. 281–286.

McGhee, T. J. (1991) Water Supply and Sewerage Engineering. 6th Edition. McGraw-Hill International Editions, Singapore.

Ng, H. Y., Elimelech, M. (2004) Influence of colloidal fouling on rejection of trace organic contaminants by reverse osmosis. *J. Memb. Sci.* **244**, 215–226.

Ng, H. Y. and Hermanowicz, S. W. (2005) Membrane bioreactor operation at short solids retention times: performance and biomass characteristics. *Water Res.* **39**, 981–992.

Nghiem, L. D. and Coleman, P. J. (2008) NF/RO filtration of the hydrophobic ionogenic compound triclosan: Transport mechanisms and the influence of membrane fouling. *Sep. Purif. Technol.* **62**, 709–716.

Nghiem, L. D., Tadkaew, N. and Sivakumar, M. (2007) Removal of trace organic contaminants by submerged membrane bioreactors 6th International Membrane Science and Technology Conference, Sydney, Australia, pp. 127–134.

Ozaki, H. and Li, H. F. (2002) Rejection of organic compounds by ultra-low pressure reverse osmosis membrane. *Water Res.* **36**, 123–130.

Pianta, R., Boller, M., Janex, M. L., Chappaz, A., Birou, B., Ponce, R. and Walther, J. L. (1998) Micro- and ultrafiltration of karstic spring water Conference on Membranes in Drinking and Industrial Water Production, Amsterdam, Netherlands, pp. 61–71.

Pronk, W., Biebow, M. and Boller, M. (2006) Electrodialysis for recovering salts from a urine solution containing micropollutants. *Environ. Sci. Technol.* **40**, 2414–2420.

Reemtsma, T., Zywicki, B., Stueber, M., Kloepfer, A. and Jekel, M. (2002) Removal of sulfur-organic polar micropollutants in a membrane bioreactor treating industrial wastewater. *Environ. Sci. Technol.* **36**, 1102–1106.

Reif, R., Suarez, S., Omil, F. and Lema, J. M. (2007) Fate of pharmaceuticals and cosmetic ingredients during the operation of a MBR treating sewage Conference of the European-Desalination-Society and Center-for-Research-and-Technology-Hellas, Halkidiki, Greece, pp. 511–517.

Sharma, V. K. and Mishra, S. K. (2006) Ferrate (VI) oxidation of ibuprofen: A kinetic study. *Environ. Chem. Lett.* **3**(4), 182–185.

Sharma, V. K. (2008) Oxidative transformations of environmental pharmaceuticals by Cl_2, ClO_2, O_3 and Fe (VI): Kinetic assessment. *Chemosphere* **73**, 1379–1386.

Stumm, W. and Morgan, J. J. (1996) Aquatic Chemistry – Chemical Equilibria and Rates in Natural Waters, 3rd Edition. Wiley Interscience, N.Y., USA.

Tan, T. W., Ng, H. Y. and Ong, S. L. (2008) Effect of mean cell residence time on the performance and microbial diversity of pre-denitrification submerged membrane bioreactors. *Chemosphere* **70**, 387–396.

Tchobanoglous, G., Burton, F. L. and Stensel, H. D. (2003) Wastewater Engineering: Treatment and Reuse. 4th Edition. McGraw Hill, N.Y., USA.

Vanoers, C. W., Vorstman, M. A. G. and Kerkhof, P. (1995) Solute rejection in the presence of a deposited layer during ultrafiltration. *Journal of Membrane Science* 107, 173–192.

Wang, D.S., Wu, X.H., Huang, L., Tang, H. X. and Qu, J. H. (2007) Nano-inorganic polymer flocculant: From theory to practice. In Chemical Water ad Wastewater Treatment IX. Edited by Hahn H.H., Hoffman E., Ódegaard H. IWA Publishing, London. pp. 181–188.

Waniek, A., Bodzek, M. and Konieczny, K. (2002) Trihalomethane removal from water using membrane processes. *Polish Journal of Environmental Studies* **11**, 171–178.

Xu, P., Drewes, J. E., Kim, T. U., Bellona, C. and Amy, G. (2006) Effect of membrane fouling on transport of organic contaminants in NF/RO membrane applications. *J. Memb. Sci.* **279**, 165–175.

Yan, M., Wang, D., Qu, J., He, W. and Chow, C. W. K. (2007) Relatively importance of hydrolyzed Al (III) species (Al_a, Al_b and Al_c) during coagulation with polyaluminum chloride: A case study with the typical micro-polluted source waters. *J. Coll. Interf. Sci.* **316**, 482–489.

Yan, M., Wang, D., Yu, J. Ni, J., Edwards, M. and Qu J. (2008) Enhanced coagulation with polyaluminum chlorides: Role of pH/alkalinity and speciation. *Chemosphere* **71**, 1665–1673.

Zhao, H., Hu, C., Liu, H., Zhao, X. and Qu, J. (2008) Role of aluminium speciation in the removal of disinfection byproduct precursors by a coagulation process. *Environ. Sci. Technol.* **42**, 5752–5758.

Zorita, S., Martensson, L. and Mathiasson L. (2009) Occurrence and removal of pharmaceuticals in a municipal sewage treatment system in the south of Sweden. *Sci. Tot. Environ.* **407**, 2760–2770.

Permissions

All chapters in this book were first published by IWA Publishing; hereby published with permission under the Creative Commons Attribution License or equivalent. Every chapter published in this book has been scrutinized by our experts. Their significance has been extensively debated. The topics covered herein carry significant information for a comprehensive understanding. They may even be implemented as practical applications or may be referred to as a beginning point for further studies.

The contributors of this book come from diverse backgrounds, making this book a truly international effort. We would like to thank all the contributing authors for lending their expertise to make the book truly unique. They have played a crucial role in the development of this book. Without their invaluable contributions this book wouldn't have been possible. They have made vital efforts to compile up to date information on the varied aspects of this subject to make this book a valuable addition to the collection of many professionals and students.

This book was conceptualized with the vision of imparting up-to-date and integrated information in this field. To ensure the same, a matchless editorial board was set up. Every individual on the board went through rigorous rounds of assessment to prove their worth. After which they invested a large part of their time researching and compiling the most relevant data for our readers.

The editorial board has been involved in producing this book since its inception. They have spent rigorous hours researching and exploring the diverse topics which have resulted in the successful publishing of this book. They have passed on their knowledge of decades through this book. To expedite this challenging task, the publisher supported the team at every step. A small team of assistant editors was also appointed to further simplify the editing procedure and attain best results for the readers.

Apart from the editorial board, the designing team has also invested a significant amount of their time in understanding the subject and creating the most relevant covers. They scrutinized every image to scout for the most suitable representation of the subject and create an appropriate cover for the book.

The publishing team has been an ardent support to the editorial, designing and production team. Their endless efforts to recruit the best for this project, has resulted in the accomplishment of this book. They are a veteran in the field of academics and their pool of knowledge is as vast as their experience in printing. Their expertise and guidance has proved useful at every step. Their uncompromising quality standards have made this book an exceptional effort. Their encouragement from time to time has been an inspiration for everyone.

The publisher and the editorial board hope that this book will prove to be a valuable piece of knowledge for students, practitioners and scholars across the globe.

Index

Printed in the USA
CPSIA information can be obtained
at www.ICGtesting.com
JSHW011400091023
49903JS00004B/36